REVISION BIOLOGY

A. E. VINES, B.Sc. (Hons.)

Head of Science Department and Principal Lecturer in Biology
The College of St. Mark and St. John, Chelsea

AND

N. REES, B.Sc. (Hons.)

Formerly Senior Lecturer in Biology
The College of St. Mark and St. John, Chelsea

By the Authors of

PLANT AND ANIMAL BIOLOGY

Volumes I and II

PITMAN PUBLISHING

First published 1963
Reprinted 1965
Reprinted 1969
Reprinted 1971
Second edition 1973
Reprinted 1975

SIR ISAAC PITMAN AND SONS LTD
Pitman House, Parker Street, Kingsway, London WC2B 5PB
P.O. Box 46038, Banda Street, Nairobi, Kenya

PITMAN PUBLISHING PTY LTD.
Pitman House, 158 Bouverie Street, Carlton, Victoria 3053, Australia

PITMAN PUBLISHING CORPORATION
6 East 43rd Street, New York, N.Y. 10017, U.S.A.

SIR ISAAC PITMAN (CANADA) LTD
495 Wellington Street West, Toronto 135, Canada

THE COPP CLARK PUBLISHING COMPANY
517 Wellington Street West, Toronto 135, Canada

ISBN: 0 273 31866 7

Reproduced and printed by photolithography and bound in
Great Britain at The Pitman Press, Bath
T.528:77

PREFACE

THE aim of this book is a very simple one. It is to assist students of biological subjects to achieve their best performances at one of the most critical tests of their careers. At the present time, the General Certificate of Education Advanced Level Examination is undoubtedly critical because upon its findings, often in terms of marks obtained in individual subjects, may depend the acceptance of a candidate for a University place. We know that the intelligent student will be aware of this and instead of proceeding hopefully without adequate preparation, will wish to come to terms with the task ahead and to plan his progress according to some predetermined pattern, during which weaknesses can be recognized and eliminated so that confidence in his own ability can grow. This book is intended to help and guide in the achievement of this confidence and to train the student to give of his best. Although prepared primarily for students at the Advanced and Scholarship levels of the General Certificate of Education, we feel that it might usefully be referred to by students at higher levels, since the biological principles embodied in it should not be lost sight of at any stage.

In the preparation of the study guide our primary consideration has been to present it as the main principles on which we think biological study at this level should be based and which, if properly understood, should serve as a basis for higher study. It is intended to incorporate a blend of the modern concepts of biological phenomena, based on the physics and chemistry of protoplasm as the living substance, and the older, but still necessary, approach based on accurate, detailed observations on whole organisms. Both are essential to the full development of the subject in the future and we see no reason why, at this level, the former should be less heavily stressed than or, as seems frequently to occur, be completely subjugated by the latter. Any observation made on a living thing, structural, physiological, reproductive or ecological, must ultimately be interpreted in terms of protoplasm and its unique properties.

We should like to thank our wives for their extremely valuable assistance during the preparation of the manuscript and during the proofing processes and to thank also our publisher for continued help and guidance.

Finally, to all those who use the book and find it of value we wish the best of good fortune in their forthcoming examinations and in their future careers.

A. E. V.

N. R.

CONTENTS

INTRODUCTION

IT is hoped that this book will prove to be of assistance to those who have reached the stage in their study of biological subjects at which the first important assessment of their progress is soon to take place.

Prior to the sitting of any examination, it is essential that the candidate should have prepared himself in such a way as to be confident of two things—

One, *that he is fully acquainted with what it is that he may be expected to have studied*, and two, *that he has adequately studied it.* Such assurance must be soundly based of course, but when properly achieved it is the real antidote to "examination nerves" which, for the most part, result from a dread of not knowing the answers.

A soundly-based self-assurance can be won only as a result of *hard work of the right kind on the part of the student.*

By the time that a student is ready to make the best use of this book, the teacher's work will have been completed, that is, to explain and demonstrate in a comprehensible way what it is that the student can reasonably be expected to know, understand and do. *The student is the only person who can learn, understand and gain proficiency in techniques on his own behalf.*

The kind of confidence mentioned above, so necessary to the student at this stage, can come only if he is satisfied that he has passed rigorous, self-imposed tests embracing all aspects of his subject matter. The repetitive reading of the same textbook or lecture notes, that so often passes for revision, has little to commend it and is boring in the extreme. It is in the relief of this condition by the provision of suitably selected exercises, that this book may serve a useful purpose. It is not a "cramming" textbook nor is it a memory training aid. It is intended to assist the student in marshalling his factual knowledge according to some logical pattern, in developing his ability to think biologically and in comprehending the more important biological concepts.

There are three main parts. The first concerns itself with what the student may be expected to have studied. This is usually referred to as the "Syllabus" and can be found in lesser or greater detail in the appropriate book of Examination Regulations. However, unless the student possesses a much fuller knowledge and understanding of his subject than is usually the case, he will seldom be able to appreciate fully the significance of what may be printed there. Here the "Syllabus" is presented as a study guide, so framed as to enable the student to form clearly in his mind a balanced picture of what his study should include.

There then follows a filling-in of the outline and in each case this is set out in the form of exercises in which *the student must do the work*. This can be carried out properly only by writing and drawing from personal knowledge after consultation of notes, textbook or teacher. In this part *the student must be honest with himself in the strictest possible way*. No profit can accrue from supplying answers directly from a book. They must be products of the student's own comprehension of what the exercise demands. It would be of assistance if a teacher or a fellow student could be persuaded to check the work for omissions and errors although this should not be necessary if the student is critically aware of the requirements of an acceptable performance. Each exercise should be completed fully and accurately and critically examined before the next is attempted. This is important since gaps may lead to increasing inability to complete succeeding exercises.

The last part is designed to give assistance in examination technique and includes examples of answers which may be expected to gain good marks.

Finally, as an appendix, is included a section containing answers to as many of the questions as can be conveniently dealt with. To answer some of the questions to the required standard, there would be need for considerably more space than can be permitted. Where this is the case, reference is made to the Volume, Chapter and Section of *Plant and Animal Biology* (Vines and Rees : Pitman) in which appropriate reading material or relevant diagrams can be found.

PART 1

A STUDY GUIDE

FOR the convenience of students and teachers, examining bodies always give an outline of the subject matter about which they expect examinees to possess knowledge and understanding. It is sometimes no more than a list of topics, presented in a disjointed fashion, advising on neither the sequence of study nor the depth to which the study should be made. Such an abstract is useful to the teacher only, who has a very wide knowledge of the subject and a good deal of experience of teaching it. The teacher can use the syllabus as a basis for a course of study in which the subject can be developed smoothly and at the right level. Teachers vary in their interpretation of what is required and not all would agree with the content and sequence of the study guide which it is proposed to outline here. Nevertheless, an attempt is made to present to the student the framework of what he should have studied in an integrated fashion and in a logical sequence. It is based on the recognition of certain fundamental truths or principles of biology as they are at present generally accepted.

The Biologist's Aims

The young biologist must, through personal observation and experiment, coupled with appropriate reading, seek to become aware of and to understand certain fundamental truths relating to living things, to become proficient in the application of as many as possible of the methods and techniques necessary to the discovery and demonstration of these principles and to become skilled in the art of communicating observations and discoveries to others. His work should be aimed at achieving an understanding of the biological concepts outlined below and it should be clear from what follows that a parallel study of the Physical Sciences is essential to the full realization of this aim.

Biological Concepts

1. *Of all systems of particles known to exist, only that known as protoplasm is "alive."*

Protoplasm is the only known system which can energize, materially sustain and replicate itself. A non-living system, lacking these properties, must ultimately disintegrate under the influence of external forces. Protoplasm passively disintegrates only when it is "dead," i.e. has lost its self-energizing, self-sustaining and self-replicating powers. The property of being alive is an expression or manifestation of these unique characteristics of protoplasm.

The student must possess a clear understanding of the meaning of "being alive."

2. Protoplasm most often occurs as units (cells), derived from similar units, and all such units exhibit the same basic structural and functional pattern.

The work of "being alive" is performed by specific arrangements of molecules under a central control. There is differentiation of molecular structure within the cell which may be related to particular functions, e.g. membranes, cytoplasmic components and nuclear structures. There may be many manifestations of cellular activity. In the performance of work there is a "power unit" common to all (ATP). A cell may possess the power of replication.

The student should possess a good up-to-date knowledge of the fundamental structure of cells as interpreted by the latest techniques of investigation, of the physical and chemical nature of the main cell constituents and the role of ATP. The mechanism and significance of cell replication (mitosis) should be studied.

3. Protoplasmic units must be in constant contact with an external inorganic system in order that the functional activities (metabolism) may be carried on.

There must be material and energy exchanges between the cell and its surroundings. The contact is maintained through an aqueous background common to cell and environment so that there is continuity between the protoplasmic material and the external inorganic world. Exchanges, for the most part, are effected through the membrane and under control of the living system which can remain functionally and structurally intact only by the expenditure of energy, since the two systems are not in a state of concentration equilibrium. When the correct contacts are established, under appropriate conditions of temperature, etc., the cell will function actively, i.e. metabolize.

The student must have a clear appreciation of the necessity for energy transfer and expenditure, of the processes by which materials enter and leave cells and the factors affecting these. The significance of the main metabolic processes in which a cell may be actively engaged should be understood, these to include the construction of "power units," synthesis of more cell substance, secretion, storage and the elimination of waste products.

4. *Most living things are composed of many cells, of varying kinds, each kind fashioned for a specific purpose and the whole co-ordinated, self-regulated system of cells, tissues, organs and organ systems, i.e. the organism, exhibits its property of being alive in a number of observable ways.*

The observable activities of material intake, gaseous exchanges, movement, growth and development, excretion, response to stimuli and reproduction by an organism are all manifestations of the co-ordinated and regulated function of a protoplasmic system.

The student must attempt to relate the external manifestations of the "life" of an organism to the structure and function of its protoplasm.

5. *In the same way that each cell has an environmental background, so has each organism, and all its functions must be performed under conditions imposed by this environment.*

An organism cannot be regarded in isolation from its environment. All its features can in some way be related to the external conditions under which it exists. It shows adaptation or is "fitted" to survive in its environment.

The student must know in some detail, the physical and chemical properties of the main classes of environments, e.g. marine, fresh-water and terrestrial, and understand the biological significance of these properties. There must be recognition of the connexion between structure and function in the organism and the physical and chemical world in which it lives, with special reference to the survival value of adaptations.

6. *At the present time we recognize an enormous variety of protoplasmic patterns in the forms of a vast number of species of plants and animals. Each kind of plant and animal is a distinct manifestation of protoplasmic activity, adapted for survival under a particular set of conditions.*

The student must possess at least an outline knowledge of this range of living things and must understand the principles upon which they are classified. Surveys of the main groups of plant and animals should be carried out on comparative bases, with special attention to increasing complexity of structure and function, structural, physiological and reproductive adaptations to environments, variation in modes of life, life cycles, survival and dispersal mechanisms. To form a framework for the survey and to obtain training in making full biological studies, certain "type" organisms should be treated in detail. These should be chosen to illustrate particular features of development, structure, function, way of life or evolutionary significance. The following should be a wide enough range from which to select—

A unicellular motile green alga.
A colonial motile green alga.
A unicellular non-motile green alga.
A colonial non-motile green alga.
A simple filamentous green alga.
A branched filamentous green alga.
A coenocytic filamentous green alga.
A diplobiontic, thalloid green alga.
A haplobiontic, thalloid brown alga.
A diplobiontic, thalloid brown alga.
A saprophytic phycomycete.
A parasitic phycomycete.
A saprophytic ascomycete.
A parasitic ascomycete.
A saprophytic basidiomycete.
A parasitic basidiomycete.
A lichen.
A liverwort and/or moss.
A homosporous pteridophyte.
A heterosporous pteridophyte.
A primitive spermatophyte.
A flowering plant.
A green flagellate.
A parasitic flagellate.
A free-living rhizopod.
A parasitic rhizopod.
A free-living ciliate.
A parasitic sporozoan.
A sponge.
A diploblastic metazoan.
A free-living, triploblastic, acoelomate metazoan.
A parasitic, triploblastic, acoelomate metazoan.
A primitive metamerically segmented, triploblastic, coelomate metazoan.
A crustacean.
A free-living insect.
A parasitic insect.
A social insect.
A lamellibranch mollusc.

A gasteropod mollusc.
An echinoderm.
A primitive chordate.
An aquatic craniate.
An amphibious craniate.
A primitive terrestrial craniate.
A winged terrestrial craniate.
A mammal.

7. *This range of form and function culminates in organisms which have reached the "highest" degree of complexity, i.e. dominate the present flora and fauna of the world, through being structurally, functionally and reproductively better able to survive in competition with others in the same environment.*

Such dominant forms in terrestrial surroundings are the flowering plants and the mammals. Partly because these represent in their own fields the present culminating points of protoplasmic development, partly because of their economic importance to man and partly because man himself is a mammal, the special characteristics of these groups should be studied in detail.

The student must gain a sound knowledge of the following aspects of angiosperm and mammalian biology—

Structure, the range of morphology, histology and anatomy with particular reference to adaptation to environment and mode of life.

Physiology, to include the study of water relationships, co-ordinating and regulating mechanisms, homeostasis and endogenous rhythms, metabolism (the concept of material and energy transformations including the nature and role of the chief organic substances and the nature and role of enzymes), nutrition (the concepts of photoautotrophism and the direct use of light energy, i.e. photophosphorylation, and heterotrophism or the inability to make use of other than the chemical energy of organic compounds), respiration (the concept of oxidative phosphorylation), locomotion (energy considerations and the mechanism of muscular contractions), growth (of populations and of the individual), development (embryology and the subsequent pattern of development over the life cycle), excretory processes (chemical and physical aspects of the elimination of waste materials), detection of and responses to stimuli, behaviour. (Note that details of the chemistry of the metabolic processes, e.g. Krebs' cycle, are not required at this level.)

Reproduction, the mechanisms and significant features of asexual and sexual reproductive processes and dispersal mechanisms.

8. *No matter how closely related a group of organisms may be, no two members of the group will ever be precisely identical in all respects. Some of the variation may be inherited.*

Variation among members of a population is always present. It differs in extent and can be traced to one of two influences, the environment or the parentage.

The student should be able to make a simple mathematical examination of variation within a population and understand how it is possible to distinguish between the influences of "nature" and "nurture."

The principles of Mendelian inheritance must be fully understood and the link between these and chromosome behaviour during meiosis. The concept of genes and the relationship between the "genetic message" in the DNA structure of the chromosomes and the translation of this into cell activity should be considered.

9. *The "higher" forms referred to above have arisen from more primitive ancestors.*

Since life appeared on earth the protoplasmic pattern has been changing.

The student should have knowledge of the range of past forms of living things, the evidence for evolution and an understanding of the processes by which it may have been brought about with reference to the origin of variation (gene and chromosome mutations) and natural selection.

10. *Organisms do not exist independently of one another but in closely integrated communities governed by the nature of the environment and the interrelationships of the community members.*

The student must take part in the making of detailed analyses of as many living communities as possible in order to compare the constitutions of naturally occurring associations and must study the physical, chemical and biotic factors which control their occurrence.

11. *World distribution of plants and animals under natural climatic conditions falls into a well-defined pattern and this pattern is in great measure the expression of world climatic and edaphic variations.*

The student should gain at least an elementary knowledge of the geography of plants and animals and should have some comprehension of the ways in which man has influenced this natural distribution.

12. *From knowledge distilled from the foregoing study, man is learning how best to control his competitors in the struggle for existence, to conserve and to make use of them for his own purposes and to widen his own environment.*

The welfare of the human population is inextricably bound up with man's relationship with other living things.

The student must gain some knowledge of the economic aspects of biological study, e.g. the recognition and control of disease-causing agents, especially the micro-organisms; the world's food problems and their solution; soil conservation and improvement; industrial uses of plants and animals; plant and animal breeding. He should become aware of the increasing danger of environmental pollution to living systems.

PART 2

TEST EXERCISES

ASSUMING that the reader has by now digested the framework of his subject, he should recognize that if he is in command of the main facts relating to each of the concepts, he should be in the comfortable position of being able to extract intelligently the relevant ones when the time comes. The important step at this stage, therefore, is to find out whether the necessary detailed information has, to put it biologically, been ingested, digested, absorbed and assimilated. There is one sure way only of knowing this, namely, to make the test by completing successfully the following part of this book. In it will be found a series of exercises, each relating mainly to the topics outlined in the study guide and cast in the form of instructions or questions most of which need only be dealt with in an abbreviated way. Before commencing any exercise, the student is advised to study his notes and textbooks until he is satisfied that the requisite information relating to the topic has been absorbed and memorized. Whenever any exercise or part of it cannot be completed, he should reacquaint himself with the necessary subject matter and repeat the test. Many of the exercises will call for ability to draw and label appropriate diagrams. Where necessary these should be drawn, since mental visualization of a diagram is seldom an adequate test of this ability. Following each series of exercises are notes concerning the practical work with which the student should be familiar, since exercises based upon it may appear in the practical examination.

Exercises marked * are considered more suitable for scholarship candidates.

The Meaning of "Being Alive"

1. Assuming that the universe is "running down," what must be the ultimate end of all inanimate matter?

2. By what unique properties can protoplasm be distinguished from inanimate matter?

3. For what fundamental purposes does the protoplasmic system require energy?

4. For what fundamental purposes does the protoplasmic system require materials?

5. What is the ultimate source of (*a*) the energy, (*b*) the materials?

6. In broad terms, how does the protoplasm perpetuate itself?

7. Under what circumstances can protoplasm be described as "dead"?

8. What is the ultimate end of "dead" protoplasm?

Cell Structure and the Nature of Protoplasm

1. Distinguish between cellular and non-cellular protoplasmic structures, giving examples of each.

2. What is a coenocyte? Give examples.

3. Give a modern definition of the term "cell."

4. What are the fundamental components of all cells?

5. Make a large clearly labelled drawing to represent the structure of a mature living cell of a plant *as seen with a light microscope*. (Include parts of adjacent cells.)

6. Write down the dimensions of the cell you have drawn.

7. How are plant cells cemented together?

8. Of what substances may plant cell walls be constructed?

9. What are the origins of (*a*) pits and (*b*) intercellular spaces?

10. Where would you look for plasmodesmata?

11. By a series of labelled drawings indicate the changes in form undergone by the cell drawn in 5 above during its differentiation.

12. Summarize these changes in words.

13. Repeat exercises 5 to 12 where applicable, for a chosen animal cell.

14. Distinguish between the living and the non-living components of cells.

15. Name and briefly describe, with functions where known, at least three living cytoplasmic components common to both plant and animal cells.

16. Repeat (*a*) at least one peculiar to plant cells and (*b*) at least one peculiar to animal cells.

17. Name and briefly describe, with purpose where known, the range of non-living inclusions which *may* be found in (*a*) plant and (*b*) animal cells. In every case state the kind of cell in which it may be found.

18. What are the main components of the nucleus?

19. What is thought to be the nature and composition of these?

20. Broadly, what are their functions and significance?

21. Summarize the main differences between plant and animal cells.

22. How may these differences be correlated with differences in body form and function?

23. What factors may be considered to determine the size of a cell?

24. How much has the electron microscope contributed to the elucidation of cell structure? What are its limitations?

25. In what fundamental ways is protoplasm a unique substance by comparison with all other molecular systems?

26. What are the main chemical constituents of actively metabolizing protoplasm? Give the approximate proportions in the general case.

27. Why could not this be applied to all cases?

28. In what physical state is most of the living substance?

29. In what way were crystalloids and colloids originally distinguished?

*30. What is the true distinction between them in terms of particle size? Compare with solvent.

*31. Explain how gold, for example, may exist in all states from insoluble to completely water soluble.

*32. What relationship is there between the gold and the solvent particles in each state you name?

*33. Distinguish between microns, sub-microns and amicrons by their dimensions. Give examples of each.

*34. What form of solution, if any, will each of these particles give?

35. By what names are the two phases of a colloidal solution described?

36. Explain the difference (use a diagram if necessary) between a colloidal sol and a colloidal gel.

37. Under what conditions may one be converted to the other? Give examples.

*38. Distinguish between suspensoid lyophobic colloids and emulsoid lyophilic colloids. Give examples.

39. Comment briefly on the following properties of colloidal solutions— light scattering, diffusibility, osmotic, electrical, stability, imbibitional, surface area and energy of the particles, exhibitors of Brownian movement.

*40. Explain how an emulsoid may stabilize a suspensoid.

41. Distinguish between absorption and adsorption.

42. Give as many reasons as you can for saying that protoplasm is in the colloidal state.

43. From what sources may protoplasm obtain its supplies of energy?

44. What appears to be the fundamental "power unit" employed in all protoplasmic work?

45. Summarize broadly the purposes for which this energy may be expended.

46. Explain how energy may be "stored" in living cells. Give both plant and animal examples.

PRACTICAL EXERCISES

1. Make temporary preparations, examine and make large, labelled drawings of selected plant and animal cells to show the fundamental structure of each.

2. Make preparations, examine and make suitable drawings of fresh material (where possible) or use permanent preparations to illustrate as many of the various cell constituents and inclusions as possible.

3. Learn to identify quickly all the main cell components.

4. Learn to "focus through" a preparation.

5. Measure the dimensions of some cells using eyepiece and stage micrometers.

6. Measure accurately the water content of fresh tissues.

7. Using chemical tests on extracts of fresh tissues demonstrate the presence of the main organic constituents.

8. Demonstrate the presence of minerals in plant ash by simple flame tests.

9. Using suitable materials such as gelatin, starch or albumin examine and demonstrate the main properties of substances in colloidal solution.

(Record all techniques, observations, measurements and conclusions in concise but accurate detail. This advice should apply to all practical exercises.)

Cell Replication

1. In terms of cells, from what is the body of every multicellular organism derived?

2. By what process?

3. What are the three main consecutive events which constitute the whole of this process?

4. Draw a series of large, clearly labelled diagrams to represent the behaviour of the nucleus and associated structures for (*a*) animal cells and (*b*) plant cells.

5. What significant differences are there between the nuclear behaviour in the two types of cells?

6. What significant similarities are there?

7. What is the value of describing stages in the process by names? What are these names and to which stages do they refer?

8. Explain, with diagrams, the formation of the middle lamella in plant cells and the subsequent formation of the primary cell wall.

9. How may pits in the cell wall originate?

10. Of what substance is the primary cell wall constructed and how are the components arranged?

11. What is a secondary cell wall, when may it be formed and of what may it be composed?

12. Briefly, using diagrams, outline the processes which follow nuclear division in animal cells.

13. At some point in the life cycle of most organisms, certain cells divide in a manner different from that which you have just described. At what stage in the life cycle of (*a*) most animals and (*b*) most plants, does this second type of division occur and what is it called?

14. Draw a series of large, clearly labelled diagrams to represent the behaviour of the nucleus and associated structures for (*a*) animal cells and (*b*) plant cells.

15. Name, in sequence, all the recognizable stages in the process.

16. Note down all the differences in appearance and behaviour of the nuclear components in the two types of division.

17. Construct diagrams to summarize the fate of the nuclear material in each of the two cases.

18. Assuming some knowledge of genetics, summarize the significance of each of the two types of cell division.

19. To what do each of the following terms refer—aster, spindle, chromosome, chromatid, centromere, chromomere, bivalent, chiasma, crossing-over, stem body, phragmoplast, cell plate, primary pit field, homologous chromosomes, haploid, diploid, reduction division?

PRACTICAL EXERCISES

1. Examine and make large, clearly labelled drawings of cells from suitable permanent preparations to show the various stages in both types of cell division. (Onion root tips, sea urchin eggs, developing spores of pteridophytes, developing pollen grains, *Ascaris* eggs.)

2. Stain onion root tips by the Feulgen method and make smears to demonstrate mitotic figures.

Material Exchanges between Cells and External Solutions

1. What is the surface layer of cytoplasm called?

2. What is the nature of this boundary, what happens to the protoplasm if it is destroyed and does it differ fundamentally (as far as is known) in plant and animal cells?

3. Despite this apparent isolation of the protoplasm, in what way is it continuous with its surroundings?

4. What must take place through the surface layer to ensure continuous activity of the enclosed protoplasm?

5. How could you demonstrate that a plant protoplast may change its volume?

6. How do you describe plant protoplasts whose volumes are much smaller than normal?

7. How could you bring them back to normal and what is the change called?

8. How could you demonstrate the comparable changes in animal cells?

9. What property of the external medium is it which influences the cells in this way?

10. By what terms do we differentiate between the movements of solute and solvent particles?

11. Is there any real difference between them and how are they affected by the presence of (*a*) a completely permeable membrane, (*b*) an impermeable membrane, and (*c*) a semi-permeable membrane?

12. Distinguish between these three classes of membranes in terms of pore and particle sizes.

13. Briefly describe a method of making a semi-permeable membrane.

14. What will be observed if such a membrane is used to separate two solutions of the same solute but of unequal concentrations?

15. Explain this observation in physical terms.

16. Define osmotic pressure as fully as you can. When does it come into existence and under what conditions may it be transient?

17. How could this pressure be measured?

18. Define the terms isotonic, hypotonic and hypertonic.

19. State the conditions which determine the osmotic pressure of a dilute solution.

20. How far can such solutions be said to obey the Gas Laws?

21. What is the OP of a solution made up from one mole of solute in 22·4 litres (dm^3) of solution?

22. What is the theoretical OP of a molar solution (1 mole solute in 1 litre of solution) of cane sugar?

23. Calculate the theoretical OP of a solution made up of 100 g cane sugar in 500 cm^3 solution.

24. What must be taken into account in calculating the OP of an electrolytic solution?

25. Calculate the theoretical OP of a solution made up of 10·1 g potassium nitrate in 100 cm^3 solution.

26. How are OPs of mixed solutions calculated?

27. Explain on a physical basis the change in appearance of plant and animal cells when immersed in solutions hypertonic to their own cell fluids.

28. In the case of plants, what part, if any, does the cell wall play in modifying the behaviour of the protoplast?

29. Explain on a physical basis the change in appearance of plant and animal cells when placed in hypotonic solutions.

30. In the case of plants, what part if any, does the cell wall play in modifying the behaviour of the protoplast?

31. Including the part played by the cell wall state the nature of *all* the forces influencing the movement of water into or out of a plant cell. What are the forces called?

32. State the name given to the resultant of all these forces.

33. Write an equation representing the relationship between these forces.

34. What is implied if the resultant has a negative value?

35. What term is used to describe a plant cell in a condition such that (*a*) this resultant is zero? (*b*) it has a maximum positive value?

36. Draw a diagram representing the changing values of all the forces mentioned as a plant cell changes in condition from one to the other of the above.

37. What can be expected to happen when living cells are continuously immersed in fresh water?

38. In the case of animals what must occur to prevent such an effect from being detrimental to the cells?

39. Why is such a process not exhibited by most plant cells?

40. In what ways is it advantageous to a plant to be in contact with a solution hypotonic to its cell fluids?

41. Mention any ways in which the penetration of dissolved substances into cells may be observed.

42. Mention any ways in which escape of dissolved substances may be observed.

*43. Distinguish between "passive" and "active" transport of materials across a cell membrane in terms of the energy expended by the cell.

*44. If "passive" movement is regarded as being equivalent to diffusion, state the laws which govern the movement of materials by this process.

*45. Can the movement of all solutes through cell membranes be regarded as "passive"?

*46. Which class of substances appears to move in this way and which properties appear to effect their rates of movement? Give named examples of such substances.

*47. Which class of solutes appears not to move "passively"?

*48. What pieces of evidence are there to support the hypothesis that the cell may be actively transporting such substances? Give examples of such experimental evidence.

*49. What bearing has the rate of metabolism of the cell on "active transport"?

*50. What is meant by the "carrier hypothesis"?

*51. How might this account for the differences between "active" and "passive" transport?

*52. Construct a simple diagram to represent the structure of the cell membrane which will allow for most of the observed permeability phenomena.

*53. Explain the meaning of the terms—antagonism, selective absorption, accumulation, physiologically balanced solution, pinocytosis, phagocytosis.

PRACTICAL EXERCISES

1. Observe and record the effects of immersing various plant cells and red blood cells in hypotonic and hypertonic solutions.

2. Make a copper ferrocyanide membrane in a porous pot and use it to demonstrate osmotic phenomena.

3. Repeat using animal membranes or plant tissues as the semipermeable membrane.

4. Make an estimate of the OP of beetroot cell sap.

5. Make an estimate of the OP of dandelion stalk cells.

6. Observe the osmoregulatory process in an amoeba or a paramecium.

7. Observe the penetration of dye (methylene blue) from a very dilute solution into living plant and animal cells.

8. Observe the effect of chloroform vapour on beetroot cells.

9. Observe the effect of the emission of CO_2 by roots on bromo-thymol-blue indicator.

10. Observe the process of diffusion of a solute through a colloidal system using gelatin and sodium hydroxide with phenolphthalein as the indicator.

11. Observe the antagonistic effects of Na^+ and Ca^+ on the permeability of beetroot cells.

Hydrogen Ion Concentration and Buffering

1. Upon what does the degree of acidity or alkalinity of a solution depend?

2. How is a neutral solution defined?

3. What is meant by total acidity?

4. Explain why the total acidity of equinormal hydrochloric and acetic acid solutions (each containing 1 g-equivalent of replaceable hydrogen) is the same, whilst their degrees of acidity are different.

5. Which of total acidity and degree of acidity seems to affect living processes?

6. Why can pure water be defined as neutral?

7. What is the value in g-ions per litre of the product of the hydrogen and the hydroxyl ions in a neutral solution at $25°C$?

8. In pure water, what will be the quantity of hydrogen ions present in g-ions per litre? Ditto hydroxyl ions?

9. Why need we use only one of these values to express both acidity and alkalinity of a solution?

10. How is the value of hydrogen ion concentration expressed in a convenient way?

11. Explain the meaning of $pH = \log 1/H^+$ or $pH = -\log H^+$.

12. State in words what this means.

13. Explain fully the meaning of a statement such as "the pH of the soil was found to be 5·8."

14. In what ways can pH be measured?

15. Define the term buffering.

16. Name environments where pH is inclined to be (*a*) very constant and (*b*) very variable.

17. What is the pH of the body fluid of a named animal?

18. What degrees of tolerance to pH changes do (*a*) animals and (*b*) plants generally tend to show?

19. Name one acid-loving and one alkali-loving plant.

*20. What aspects of cell metabolism might be affected by pH variations? Give possible reasons.

*21. How may pH changes affect intake of materials by cells?

PRACTICAL EXERCISES

1. Using the appropriate test papers or indicators measure the pH of tap water, sea water, samples of pond water, expressed juices of tissues, e.g. orange juice, and soil samples.

2. Observe the buffering effect of an ammonium acetate solution.

Energy Exchanges

1. From an energy intake and utilization viewpoint, compare and contrast a living cell with a machine.

2. In what form does the energy used by all living cells appear to occur within the cells?

3. What may be the original source of this energy?

4. Which classes of organisms can tap such a source directly?

5. How do the organisms which have no direct access to such natural sources of energy obtain their supplies?

6. Give at least four ways in which energy may be dissipated to the external surroundings by living organisms.

7. Give at least three purposes for which energy may be used internally by living organisms to do the work of the protoplasm.

The Nature of the Chief Constituents of Protoplasm and the Products of its Metabolism

1. Of what substance is the bulk of a living organism composed?

2. Give at least six of its properties which have significance in the structure and functions of protoplasm. In each case state what this significance is.

3. What three main groups of organic substances make up the bulk of the remaining protoplasmic material?

4. For each of these—(*a*) Make a classification of the commoner kinds of compounds within the main group based on their chemical nature. (*b*) Give specific examples of substances in every kind of compound so distinguished. (*c*) Against each substance named say where it might be expected to be found and as far as possible what role in the life of the organism it may be playing.

5. Define a sugar.

6. Distinguish between—pentose and hexose, aldose and ketose, reducing and non-reducing, mono-, di- and polysaccharides, pentosans and hexosans.

7. Give examples of each of these.

8. What sugar units may be derived from maltose, sucrose, lactose, cellulose, starch, glycogen, inulin, stachyose, raffinose, xylan, araban?

9. In what ways may these substances be broken down into their constituent sugars?

*10. Briefly outline the chemical nature of hemicellulose, pectin, mucilage, gum, chitin, amygdalin, mustard oil, digitalin, plant sap pigments, mannitol, glucosamine.

*11. Give examples of where these substances may be found.

12. Define an ester and give an example.

13. Of what two classes of compounds are fatty substances composed? Which of these is common to all simple fats? In the case of the other, name as many examples as possible.

*14. Show chemically how any one named fat may be formed from its two primary constituents.

15. Name some fats and oils found in plants and animals and say from which parts they may be extracted.

16. Distinguish between a fat and a wax. Give some examples of the latter and indicate their sources.

17. How may fatty substances be broken down to their main constituents?

*18. Briefly outline the chemical nature of lecithin, glycocholic acid, calciferol, cholesterol. Where may these substances be found?

19. Define an amino acid.

20. Name at least ten amino acids commonly found in plants and animals. About how many more are known to occur naturally?

21. Why are amino acids referred to as amphoteric electrolytes?

*22. What is meant by the iso-electric point?

*23. What is a zwitterion?

24. Show chemically how two named amino acids may be united.

25. What name is given to such a chemical union?

26. What kind of substance is formed from the union of (a) two amino acids? (b) many amino acids?

27. How may chains of amino acids be united to form still more complex compounds? What are such compounds called?

28. In what ways may these be reduced to their amino acid units?

29. Name the kinds of compounds formed in the breakdown process.

30. Explain why these very complex substances occur in such a wide variety.

*31. What, briefly, is the chemical nature of albumin, keratin, legumin, glutelin, caseinogen, nucleoprotein, haemoglobin, phycocyanin, cytochrome, enzymes, insulin?

*32. Where may such substances be found and what purposes may they be serving there?

33. Name about eighteen chemical elements which seem to be present in almost every living thing.

34. State what appears to be the chief part played by each of these.

PRACTICAL EXERCISES

1. Using reasonably pure examples of the important organic compounds featured above, learn to detect their presence by the appropriate chemical tests.

Details of at least the following should be recorded—Molisch's test, Barfoed's test, Benedict's test, Fehling's test, the Iodine test, the Grease-spot test, the osmic acid and Sudan III tests, the xanthoproteic test and Millon's test.

Enzymes and their Role in Metabolism

1. Fully define an enzyme, being careful to distinguish it from any inorganic substance which may possess comparable properties.

2. Using examples, give the ways in which most enzyme names are chosen.

3. What is meant by the "specificity" of an enzyme?

4. Give examples to show what main forms this specificity may take.

5. Name at least three factors which are known to influence the activity of enzymes. In each case say in what way the influence is apparent and if possible say why. Quote actual examples of enzymes in these connexions.

6. Distinguish carefully between apoenzymes, coenzymes, holoenzymes, activators, inhibitors and anti-enzymes, giving examples.

7. Name four classes of hydrolysing enzymes, stating broadly in each case the kinds of reactions in which they are involved. Give the names of at least one plant and one animal enzyme in the classes named, with the reaction.

8. Name three classes of oxidizing and reducing enzymes and supply details as above.

9. Name any other main groups of enzymes and supply the details.

10. Which reactions are associated with the following enzymes—ptyalin, carbonic anhydrase, catalase, invertase, lactic dehydrogenase, pepsin, maltase, starch phosphorylase, zymase?

11. Explain why it is necessary to run a boiled control when attempting to demonstrate the activity of an enzyme.

PRACTICAL EXERCISES

1. Demonstrate the activity of the following enzymes on suitable substrates—lipase, invertase, amylase, pepsin, zymase, oxidase, peroxidase, catalase, rennin, dehydrogenase.

2. Make an enzyme extract from yeast and test its activity.

3. Check the activity of invertase on sucrose using a polarimeter.

The Organism as a Co-ordinated Protoplasmic System

1. We may distinguish a living thing from a non-living structure by certain characteristics peculiar to the former. What are these? In other words, what are the main external manifestations of "life."

2. For each that you name broadly establish its connexion with protoplasmic activity.

3. Define the terms cell, organelle, tissue, organ, organ system and organism.

Environmental Conditions

1. Define the term environment.

2. Distinguish between internal and external environments.

3. On a broad basis distinguish between the main external environments of living things.

4. Can these be as sharply divided as your distinction between them suggests?

5. If not, enlarge on this by giving transitional cases.

6. Give at least six physical properties of sea-water which have some biological significance. In each case say what this may be quoting the relevant data.

7. Repeat with the chemical properties.

8. Repeat with the physical and chemical properties of fresh water.

9. What two main components must be taken into account when considering the terrestrial environment?

10. List the components of a fertile soil.

11. Summarize the facts about the composition of each of these.

12. Give at least seven properties of a soil which may be considered to have some influence on its inhabitants.

13. Against each say in what ways these influences may be apparent.

14. Briefly classify soils on a particle size basis.

15. List the special characteristics of (*a*) sandy and (*b*) clayey soils.

16. How may each of these be improved and what is the effect of the treatment in each case?

17. Define the terms soil horizon, weathering crust, mull, mor, marl, loam, brown earth, podsol, rendzina.

18. Give at least seven properties of the atmosphere which may be of biological significance. In each case say what this may be quoting the relevant data.

19. What is meant by organic environments?

20. Summarize the chief ways in which aquatic and terrestrial environments differ.

PRACTICAL EXERCISES

1. Measure the specific gravity and pH of sea-water and fresh-water from various sources.

2. Measure the chloride content of sea-water and estimate its other ionic constituents.

3. Estimate the quantities of dissolved gases in sea-water and fresh-water from various sources.

4. Make mechanical analyses of various soils to study their texture.

5. Measure the organic and water contents, sticky points, lime deficiencies and pH of various soils.

6. Compare the capillarity and drainage properties of pure sand and pure clay.

7. Dig a suitable trench and study a soil profile.

The Classification of Living Things

1. What is meant by each of the terms taxonomy, systematics and nomenclature?

2. Name the seven major categories of classification into which organisms may be placed.

3. What is meant by the binomial system of nomenclature? Who devised it?

4. Which name must always be written with a capital letter?

5. What other rules do you know which must be obeyed in the naming of plants and animals?

6. What taxonomic methods have been used in the compilation of present day classifications?

7. Which of these are the more recent?

8. Which methods have been most widely used and why?

9. What is meant by the terms phylogenetic relationships, species, *nomina conservanda*?

10. Name the major divisions of the plant kingdom.

11. Break each into its main classes.

12. Name at least ten phyla of the animal kingdom and break each into its main classes.

13. Write down in ascending order the full Latin names of (*a*) the plant types and (*b*) the animal types which you have studied and place each in its appropriate division or phylum.

14. What is meant by a "natural" classification?

15. Why may this never be achieved in the fullest sense?

THE PLANT KINGDOM

Algal Forms

1. What are the main distinguishing features of algae?

TYPE STUDIES

2. Write down the generic and, if possible, the specific names of the algae which you have studied.

3. State the general environmental conditions under which each exists with reference to such factors as light availability, water availability, nutrient sources, oxygen supply, body support, motility and any others which may have influence on the plant's survival.

In the case of each type named—

4. Make large, clearly labelled drawings to represent the development and mature vegetative structure of the plant, including features of external morphology, tissue disposition and cell detail as necessary and appropriate. Give indication of sizes.

5. State the presumed function of each organ, tissue, cell or cell component which you illustrate.

6. Note any structural and functional adaptations to environment which are apparent, e.g. light-catching parts, specialized absorbing structures locomotor apparatus, organs or cells of attachment, supporting structures osmoregulatory apparatus.

7. State the "mode of life," source of nutritional substances and how these are obtained and utilized.

8. Show, with diagrams, how growth is manifested.

9. Where applicable, show how locomotion is effected.

10. State which tactic and/or tropic responses are known to be made and against each say any biological advantages which may be gained from the response.

11. Make large, clearly labelled drawings to represent any asexual reproductive structures and state the part they play in the reproductive processes.

12. Repeat for sexual reproduction.

13. Note any reproductive adaptations to environment, e.g. motility or otherwise of spores and gametes, spore and gamete dispersal mechanisms, survival during adverse conditions.

14. Note any methods of vegetative propagation.

15. Make a large, clear diagram to represent the life cycle, marking in all the important stages.

16. Note any features of economic importance to man.

17. Note any features of significance in the lives of other organisms.

GENERAL STUDIES

18. Give any structural, physiological and reproductive features which algae have in common.

19. Place the genera you have named in their divisions and distinguish briefly between these.

20. Name at least one representative of as many of the following algal forms as you can—unicellular motile; colonial motile; unicellular non-motile; colonial non-motile; simple filamentous; branched filamentous; coenocytic; expanded thallus; *mono-axial; *multi-axial; complex thallus showing tissue differentiation.

21. Which may be regarded as the most primitive form and why?

22. How far is "division of labour" exhibited by the algae?

23. How far can a unicellular alga be compared with (*a*) a whole higher plant and (*b*) a single cell of a higher plant?

24. What is the mode of nutrition in the algae?

25. Name three different food storage substances and three different pigments found in the algae and the algae in which they may be found.

26. Describe three different ways by which growth may be manifested in the algae.

27. How far is motility a feature of the algae?

28. To what variations in external conditions are algae known to make responses and what forms do the responses take?

29. Name any methods of vegetative reproduction exhibited by algae, giving examples.

30. Name at least two kinds of asexual spores formed by algae, giving examples.

31. Distinguish between isogamy, anisogamy (heterogamy) and oogamy and give examples of algae exhibiting these forms of sexual reproduction.

32. Outline the advantages of oogamy.

33. Outline the life cycle of one haplobiontic alga (not exhibiting alternation of generations).

34. Outline the life cycle of one diplobiontic alga.

35. Summarize what is meant by alternation of generations with reference to the algae.

36. What evidence is there for the existence of heterothallism in the algae?

37. Summarize the algal characters associated with the aquatic environment under structural, physiological and reproductive headings.

PRACTICAL EXERCISES

1. Collect (preferably) or obtain from a dealer fresh specimens of the types required. When collecting, note the date and place of finding and the conditions prevailing in the habitat.

2. Examine with hand lens and microscope where applicable and make large, clearly labelled drawings to illustrate in as much detail as possible all vegetative and reproductive structures.

3. Make use of permanent preparations to show structures not found or not easily demonstrated with fresh material.

4. Introduce fresh water algae into an aquarium or school pond and examine periodically for various stages in life cycles.

5. Investigate the incidence of phototactic responses by motile green algae,

6. Make a study of the zoning of sea weeds.

Fungal Forms
TYPE STUDIES

Repeat 1–17 as for algae, pp. 19–20.

GENERAL STUDIES

18. Name any structural, physiological and reproductive features which the fungi have in common.

19. What characteristics distinguish fungi from all other plants?

20. Place the genera you have named in their classes and distinguish briefly between these classes.

21. Outline the range in structure to be found among the fungi you have studied, giving examples.

22. Can all unicellular fungi be considered as primitive forms? If not, say why.

23. How does the mode of life of a fungus compare with (a) a green plant and (b) an animal?

24. In broad terms what are the nutritional requirements of fungi and how may they be satisfied?

25. Distinguish clearly between parasites and saprophytes, giving named examples.

26. Note the special adaptations, structural, physiological and reproductive of (*a*) a saprophytic and, (*b*) a parasitic fungus, using named examples.

27. How may growth be manifested in the fungi?

28. How far is motility a feature of the fungi?

29. To what variations in external conditions are fungi known to make responses, what forms do the responses take and of what advantage or survival value are they?

30. Describe the range in spore forms found in the fungi, with named examples, and state the origin of the spores mentioned.

31. How far are (*a*) asexual processes and (*b*) sexual processes in fungi adapted to the terrestrial environment as compared with algal adaptations to aquatic conditions?

32. Describe at least three different methods of spore dispersal by fungi, giving named examples.

*33. Trace as far as possible the decline of sexuality in the fungi.

34. How is heterothallism evident in any one named fungus?

35. What part is played by fungal saprophytes in (*a*) nature and (*b*) man's economy?

36. What part is played by fungal parasites in (*a*) nature and (*b*) man's economy?

37. What is a mycorrhizal association? Give some examples.

38. Survey the control measures which may be taken against (*a*) saprophytes and (*b*) parasites.

39. State at least three economic uses of fungi, giving named examples.

PRACTICAL EXERCISES

1. Using suitable culture media such as waste bread, porridge, fruits and vegetables await development of saprophytic fungi. Examine and identify as far as possible. (*Mucor*, *Rhizopus* and *Eurotium* are amongst the most likely to be obtained in this way.) Note appearance and colour of the mycelium.

2. Parasitic fungi may be collected at certain times of the year but are difficult to culture. It may be found necessary to obtain preserved specimens through an agency if the host plant does not grow locally. *Pythium sp.* will almost always develop on overcrowded cress seedlings in very moist conditions.

3. Examine with hand lens and microscope where applicable and make large, clearly labelled drawings to illustrate in as much detail as possible all vegetative and reproductive structures. Make use of permanent preparations to show structures not found or not easily demonstrated with fresh material.

4. Cover some fresh, moist horse dung and watch the succession of fungi over several months. Look particularly for *Pilobolus* and study its method of spore dispersal. Look also for *Coprinus* and make spore prints.

5. Make up sterile media (prune agar, potato agar, etc.) and inoculate with spores of various saprophytes.

6. Try to develop zygospores of *Mucor* (using pure + and − strains).

7. Make up a suitable culture medium (Pasteur's) for yeast and investigate its fermentation activities.

8. Investigate tropic responses made by *Mucor*, *Pilobolus* and *Coprinus*.

GENERAL STUDIES

Lichens

1. Write down the generic and, if possible, the specific names of any lichens you have encountered.

2. State the general characters of the environments in which they were found.

3. Of which two plant forms are lichen bodies composed? Give the names of some examples of these.

4. Describe three morphologically different lichens.

5. Make large, clearly labelled drawings of at least one of these to show external morphology and internal structure.

6. Distinguish between heteromerous and homoiomerous internal structures.

7. Illustrate how one named lichen is structurally adapted to its environment.

8. What are its nutritional requirements and how are these obtained?

9. How is growth of the lichen body manifested?

10. What physiological adaptations to its environment does it show?

11. How does it reproduce asexually and how are the reproductive parts dispersed?

12. What sexual reproductive processes are exhibited?

13. How is the association between the components of a lichen body described?

14. What are the special features of such an association?

15. In the case of the lichen how do these features apply?

16. What is the part played by lichens in the economy of nature?

17. Give at least two of man's economic uses of lichens quoting named examples.

PRACTICAL EXERCISES

1–3. As 1, 2 and 3 for algae, p. 21.

GENERAL STUDIES

Bacteria

*1. Distinguish between the true bacteria and other micro-organisms often classified with them.

2. Briefly survey the range of habitats in which bacteria may be found.

3. Give, with examples, the range of forms (shapes) and sizes of bacterial cells.

4. How does the bacterial cell compare with the generalized plant cell with respect to (*a*) wall structure, (*b*) cytoplasmic inclusions, (*c*) nuclear structure, (*d*) presence of pigments?

5. How much is motility a feature of bacterial cells and how is it effected?

6. Distinguish between photoautotrophic, chemoautotrophic and heterotrophic bacteria, giving examples of each kind.

7. How do the chlorophyll-containing bacteria compare nutritionally with green plants? What kinds of nutrients do they require?

8. Describe four main groups of heterotrophic bacteria in terms of their nutritional requirements.

9. Distinguish between aerobic and anaerobic bacteria, with named examples, and between obligate and facultative forms cf these.

10. How is bacterial growth manifested and under what conditions is optimum growth made?

11. Describe at least one response made by a bacterium to an external stimulus.

12. How is reproduction effected by bacteria?

13. Give an indication of how the size of a bacterial population may grow given the best conditions.

14. Give one way in which some named bacteria may survive adverse conditions.

15. Outline the part played by named examples of bacteria in (*a*) soil fertility (reference to putrefactive activity, cycles of nitrogen, sulphur, phosphorus, carbon, etc., humus formation, soil structure), (*b*) pathogenic conditions of plants and animals (reference to virulence, toxicity, host defence mechanisms, immunity), (*c*) symbiotic associations.

16. Outline the part played by named examples of bacteria in man's economy with reference to, (*a*) industrial uses, (*b*) destruction of man's materials and stores.

PRACTICAL EXERCISES

1. Obtain specimens of *Bacillus subtilis* by making a hay infusion. (Place a small quantity of dry hay in a beaker, cover with water and bring to the boil. Allow to stand for several days in a warm place.) Examine a drop of the fresh liquid for the motile bacilli using a $\frac{1}{12}$ in. immersion lens if available. When a scum appears on the surface of the infusion examine it for spores.

2. Make a smear of the infusion liquid and stain with methylene blue or carbol fuchsin. Find the right technique for doing this.

3. Obtain specimens of *Bacillus mesentericus* and *Clostridium butyricum* by making a potato infusion. (Place a slice of potato in a beaker and fill with water. Allow to stand for several days in a warm room.) Examine the surface water for *B. mesentericus* and water from deep in the beaker for *C. butyricum*. Why do they grow separately at different levels in the water? Check your supposition by making stab cultures in sterile potato-agar tubes.

4. Make up suitable nutrient agar plates (sterile) and check the presence or absence of bacteria in air, in unboiled tap water, boiled tap water, on finger tips, on coins, in soil, in milk, etc., by inoculation from these sources.

5. Examine permanent preparations of various pathogenic organisms using the $\frac{1}{12}$ in. oil immersion lens if available.

6. Try to arrange a visit to the local Public Health Laboratories or a Pathology Laboratory at a Hospital.

The Bryophyta

1. What are the main distinguishing features of this division?

TYPE STUDIES

Repeat 2 and 3, p. 19, for the genera studied, and for each of these repeat 4 to 17, pp. 19–20 for (i) the gametophyte and (ii) the sporophyte.

GENERAL STUDIES

18. Give the Latin names of the two main bryophyte classes.

19. Give the main features by which members of the two classes can be distinguished.

20. Which may be considered to be the more primitive of the two gametophyte forms? Give reasons. Repeat for the sporophyte forms.

21. Outline the structural, physiological and reproductive features of bryophytes which can be regarded as adaptations to terrestrial conditions.

22. How far can they be considered as fully land-adapted? Give instances of features which may be more closely related to aquatic conditions.

23. To what extent is "division of labour" exhibited by bryophytes?

24. State clearly what is meant by "alternation of generations."

25. Indicate how bryophytes illustrate this kind of life cycle.

*26. Give a reasoned suggestion as to the possible origin of the bryophytes.

PRACTICAL EXERCISES

1–3. As 1, 2 and 3 p. 21.

4. Germinate spores and gemmae (where applicable) on the surface of a moist plaster block or on moist compost in a covered pot and watch the developmental stages.

5. Watch the hygroscopic movements of elaters and peristome teeth.

The Pteridophyta

1. What are the main distinguishing features of the division?

TYPE STUDIES

As for the Bryophyta, above, for (i) the sporophyte and (ii) the gametophyte.

GENERAL STUDIES

18. Give the Latin names of the classes to which your pteridophyte types belong.

19. Give at least two features which all pteridophytes have in common.

20. Give at least two features by which they can be distinguished from (a) algal forms and (b) bryophytes.

21. If representatives of more than one class have been studied show how these classes can be separated.

22. Outline the structural, physiological and reproductive features of both sporophytes and gametophytes which can be regarded as adaptations to terrestrial conditions.

23. How far can the pteridophytes be regarded as fully land-adapted? Give examples of features clearly not land adaptations.

24. To what extent is tissue differentiation evident in pteridophyte sporophytes and gametophytes?

25. What are the main types of tissues encountered and what are their functions?

26. Indicate clearly how pteridophytes exhibit alternation of generations and compare this with the same condition in bryophytes.

27. Distinguish between homospory and heterospory, giving examples.

*28. In what way may heterospory be related to the seed habit?

29. Why are pteridophytes sometimes referred to as "vascular cryptogams"?

30. Indicate that you know the meaning of (*a*) apospory and, (*b*) apogamy, with reference to a pteridophyte.

PRACTICAL EXERCISES

1–3. As 1, 2 and 3 p. 21.

4. Sow ripe fern spores on the surface of some damp compost, cover with a glass sheet and place in a shady part of the laboratory. Observe from time to time for the development of gametophytes.

5. Make a hanging-drop culture of spores in water to see the early stages of germination.

6. Watch ripe, dry fern sporangia under the microscope to observe the spore discharge mechanism. Warm the slide if necessary.

The Spermatophyta

1. What are the main distinguishing features of this division?

A. Gymnospermae

TYPE STUDIES

As for the Pteridophyta, p. 25.

GENERAL STUDIES

18. Name the orders to which the genera you have studied belong.

19. Give at least three features which all gymnosperms have in common.

20. Give at least three characteristics by which they differ from pteridophytes.

21. If several gymnosperm orders have been studied give the chief ways in which they differ.

22. Outline the structural, physiological and reproductive characteristics of gymnosperm sporophytes which can be related to the conditions prevailing in a terrestrial habitat.

23. To what extent is tissue differentiation evident in a named gymnosperm sporophyte? What tissues are encountered and what are their functions?

24. In what fundamental ways does the female gametophyte of a gymnosperm differ from the corresponding structure in pteridophytes?

25. What advantages may be derived from these differences?

26. To what pteridophyte structure does the gymnosperm ovule correspond?

27. What structure corresponds with a *Selaginella* megaspore?

28. To what pteridophyte structure does the gymnosperm pollen grain correspond?

29. How far can it be regarded as a reduced structure?

30. Distinguish between zoidogamy and siphonogamy, giving examples.

*31. Give a full description of a seed in terms of its origin.

32. Indicate clearly how gymnosperms exhibit alternation of generations.

33. Compare this with the corresponding condition in pteridophytes.

34. Outline the advantages of the seed habit.

35. Why do you think it is useful to study a gymnosperm?

PRACTICAL EXERCISES

1–3. As 1, 2 and 3 p. 21.

4. Germinate a variety of gymnosperm seeds to watch seedling development.

B. Angiospermae

Vegetative morphology

1. Broadly, what should be included under the heading "morphology"?

2. Giving examples, distinguish between mesophytes, hydrophytes and xerophytes.

3. Draw a large, clearly labelled diagram to represent the body form of a generalized mesophytic flowering plant.

4. Make a drawing from a *named specimen* of a plant which fits this body form.

5. Distinguish between herbs, shrubs, and trees; deciduous and evergreen; annual, biennial, perennial, and ephemeral.

6. What is the origin of all sexually produced angiosperms?

7. Draw a large, clearly labelled diagram to show the essential parts of an angiosperm seed.

8. To what do the terms hilum, aril, caruncle and micropyle refer?

9. Classify seeds according to structure and give named examples of each kind.

10. Distinguish between epigeal and hypogeal germination.

11. Distinguish between the primary and secondary bodies of a plant.

12. From which part of the embryo does the shoot system develop?

13. Distinguish morphologically between a shoot system and a root system.

14. What characteristics of a stem would you include in giving its morphological description?

15. Distinguish between dichotomous and axillary branching; monopodial and sympodial branching; give examples in all cases.

16. What is the origin of a branch and in what ways may it subsequently develop?

17. What is the fundamental function of a stem?

18. What is meant by the term "modified"?

19. Survey the main modifications of stems in terms of structural adaptations to particular functions.

20. How would you distinguish a leaf from other vegetative parts?

21. What characteristics of a leaf would you look for in order to describe it morphologically?

22. What is meant by heterophylly? Give named examples of its incidence.

23. What is meant by phyllotaxis?

24. Give three different forms of leaf arrangement with named examples.

25. What is the fundamental function of a foliage leaf?

26. How is it morphologically suited to this function?

27. Survey the main modifications of leaves in terms of structural adaptations to particular functions.

28. What are the morphological characteristics of a vegetative bud?

29. How is a resting winter bud usually constructed?

30. How can a root system be defined morphologically?

31. Distinguish between a tap root system and a fibrous root system in terms of their origin and development.

32. Distinguish between intensive and extensive root systems.

33. What are the main functions of roots and how are they morphologically adapted to these?

34. Survey the main modifications of roots in terms of structural adaptations to particular functions.

35. Outline the main morphological differences between mesophytes and (a) hydrophytes and, (b) xerophytes.

PRACTICAL EXERCISES

1. Make large, clearly labelled drawings of selected mesophyte, hydrophyte and xerophyte plants to show their general form and to bring out the fundamental morphological differences between them.

2. Dissect and make drawings of a suitable variety of seeds.

3. Germinate as many different seeds as possible and record by drawings the morphological changes observed during seedling development.

4. Make large, clearly labelled drawings of suitable specimens to show the distinction between monopodial and sympodial branching.

5. Dissect, draw and clearly label, suitable named examples of—cladode, phylloclade, winged stem, rhizome, root stock, corm, bulb, turion, runner (stolon), sucker, offset, stem tuber, dropper, bulbil, twining stem, branch tendril, thorn, prickle.

6. Draw and label the main parts of leaves which show the four fundamental lamina compositions.

7. Examine and draw suitable examples of plants which exhibit heterophylly.

8. Construct phyllotaxis plans for named plants based on observed leaf divergences.

9. Draw and label suitable named examples of—scale leaf, bud scale, bract, bracteole, stipule, ligule, phyllode, leaf spine, leaf tendril.

10. Dissect and make a clearly labelled drawing to show the structure of a large bud (brussels sprout).

11. Examine and draw a variety of winter buds. Section or dissect them to show vernation.

12. Make large, clearly labelled drawings of named tap root and fibrous root systems.

13. Draw and label suitable named examples of—swollen tap root, root tuber, prop roots, aerial roots, climbing roots, "coralloid" roots, haustorial roots.

Histology

1. Define the term histology.

2. Distinguish between organism, organ system, organ, tissue and cell.

3. Make a large, clearly labelled diagram, showing only the fundamental components, to illustrate the structure of a plant cell which has just been formed at an apical growing point.

4. Such a cell might mature into one of many final forms. Say what these are. Group them into the four main types of plant tissues and say broadly what the functions of these main types of tissues are. What is meant by a compound tissue? Give examples.

5. What is the pattern of differentiation followed by a young cell? Illustrate by labelled diagrams.

6. Draw large, clearly labelled drawings to illustrate the structure of all the mature cell forms listed in 4. State how the structure of each can be correlated with its presumed function.

7. Draw a series of labelled diagrams to illustrate the stages of differentiation of a xylem vessel.

8. Distinguish between primary and secondary walls; simple and bordered pits.

9. Illustrate the variety of structure which may be observed in the secondary walls of xylem elements.

10. Draw a series of labelled diagrams to illustrate the stages of differentiation of a phloem sieve tube and companion cell.

11. For each main type of tissue listed in 4 show where such tissues may generally be found in (*a*) stems, (*b*) roots, (*c*) leaves.

12. In each case relate position to functional efficiency.

PRACTICAL EXERCISES

1. From macerated tissues (preferably) or from hand-cut or permanent preparations of transverse, longitudinal and tangential sections of plant organs observe, draw and label examples of all the different cell forms. Remember that the structure of a cell only becomes apparent when it is seen in three dimensions. Drawings of cells as seen cut in one plane only are of little value in indicating their real shape and form.

Vegetative Anatomy

1. Define the terms ontogeny and anatomy.

2. Distinguish between the primary and secondary bodies of a plant.

3. From what does the mature angiosperm plant develop? What are the fundamental parts of this?

4. What is a meristem? Where are they to be found in the structure named in 3.

5. Make a large, clearly labelled diagram to illustrate the structure of a dicot. stem apex.

6. What form does the apical meristem take?

7. What is a histogen? Name those which can be distinguished at a stem apex.

8. What other structures in addition to the stem are initiated at the stem apex?

9. Draw a large, clearly labelled diagram to represent the anatomy of a named primary dicot. stem as seen in T.S. through an internode. Distinguish clearly the epidermal, cortical and stelar regions.

10. What form does the stele take?

*11. Make labelled diagrams to illustrate the range in form of vascular bundles. Give examples of each.

12. Make large, clearly labelled diagrams to represent the origin and differentiation of a vascular bundle from the stem apex.

13. Distinguish between procambium and cambium, protoxylem and primary metaxylem, protophloem and primary metaphloem.

14. Distinguish between the endarch and exarch positions.

15. What marks the outermost layer of the stele?

16. What is the central portion of the stele called?

17. What may mark the innermost layer of the cortex?

18. What is a starch sheath?

*19. Make a large, clearly labelled diagram to represent the anatomy of a named primary dicot. stem as seen in T.S. through a node.

*20. What is (a) a leaf gap and, (b) a branch gap?

21. Make a large, clearly labelled diagram to illustrate the structure of a monocot. stem apex.

22. What form does the apical meristem take?

23. Draw a large, clearly labelled diagram to represent the anatomy of a named primary monocot. stem as seen in T.S. through an internode. Distinguish between the epidermal, cortical and stelar regions.

24. What form does the stele take?

25. Give at least three ways by which a monocot. stem may be distinguished from a dicot. stem.

26. Define clearly what is meant by secondary growth.

27. Does it occur in all flowering plants?

28. What are (a) lateral meristems and (b) intercalary meristems?

29. Make large, clearly labelled diagrams to illustrate their position in the plant.

30. Make a series of suitable diagrams to represent the stages in the process of secondary thickening in a dicot. woody stem as seen in (a) T.S. and (b) L.S.

31. What are intra- and inter-fascicular cambia? What is their origin?

32. What are fusiform initials and ray initials? What tissues are derived from these?

33. Make a series of diagrams to represent cambial activity during the formation of secondary vascular tissues.

34. What are secondary medullary rays and what is their origin?

35. What is a cork cambium, where may it arise and what tissues may be formed by it?

36. What may be the fate of a primary epidermis in a stem undergoing secondary thickening?

37. With suitable diagrams explain the formation of a lenticel.

38. Draw a large, clearly labelled diagram to represent the anatomy of a three year old woody stem as seen in T.S., R.L.S. and T.L.S. How would you know its age?

39. Distinguish between heart wood and sap wood.

40. Explain the appearance of grain and knots in a plank of wood.

41. What differences are there between hard woods and soft woods?

42. In what ways may monocot. stems increase in girth?

43. What is a primary thickening meristem?

44. Make a series of suitable diagrams to represent the stages in the process of secondary thickening in a named monocot. stem as seen in T.S.

45. What is the origin of the cambium and in what manner are the secondary tissues formed?

46. Briefly describe the ontogeny of a dicot. secondary branch.

47. Describe the appearance of a named abscissed twig.

48. Make a large, clearly labelled diagram to represent the anatomy of a dorsi-ventral leaf of a named dicot. mesophyte.

49. Point out those features which are clearly adaptations to function.

50. What differences are there between the upper and lower surface layers?

51. Make drawings to represent the structure of a stomate on this leaf. Surface and T.S. views.

52. Repeat for any other stomatal forms that may have been encountered.

53. Survey the range of epidermal outgrowths which may be found on leaves, giving named examples.

54. Indicate with diagrams how the main and subsidiary vascular tissues of leaves may be arranged.

*55. Make a large, clearly labelled drawing to represent the arrangement of tissues in a named dicot. petiole.

*56. How is the arrangement conducive to sturdiness?

57. Make a large, clearly labelled diagram to represent the anatomy of an isobilateral leaf of a named monocot. mesophyte.

58. Point out the ways in which it varies from the typical dorsi-ventral leaf.

*59. Make a series of labelled drawings to represent the development of a dicot. leaf from a leaf buttress.

60. How may the growth of a monocot. leaf differ from that of a typical dicot. leaf?

61. Make a series of labelled drawings to represent the process of leaf abscission in a general case.

62. Make a large, clearly labelled diagram to represent the structure of a root apex as seen in L.S.

63. What are the four main tissue systems to be found in a root and what names are given to the histogens from which they arise?

64. Describe the structure and function of a root cap.

65. **Make** a large, clearly labelled diagram to represent the anatomy of a primary root of a named dicot. as seen in T.S. through the root hair region.

66. Repeat for L.S. indicating the origin of the tissues from the apical meristem.

67. What is meant by exarch protoxylem, pentarch root and piliferous layer?

68. Summarize the differences in anatomy between the primary roots of dicots. and monocots. in the general cases.

69. Make a series of suitable diagrams to represent the stages in the process of secondary thickening in a dicot. woody root.

70. In what ways does the process differ from the same process in a stem?

71. Draw a series of diagrams to illustrate the development of a lateral root.

72. How does this differ from the development of a lateral shoot?

73. With named examples and suitable diagrams summarize the main anatomical characteristics of xerophyte and hydrophyte plant organs. Point out where the anatomical features are clearly adaptations to environment.

PRACTICAL EXERCISES

1. Cut, suitably stain, mount and examine transverse and longitudinal sections of the major vegetative organs of named examples of monocots. and dicots. using appropriately fixed and hardened material. Temporary preparations are usually adequate and the student should become familiar with the use of Schultze's solution (chlor-zinc-iodide), aniline sulphate, aniline chloride and acidified phloroglucin as suitable staining reagents. If it is desired to make permanent preparations, the student is advised to gain experience in the use of safranin, Delafield's haematoxylin and light green in clove oil.

In all possible cases illustrate the anatomy of the part concerned by (i) labelled plans of tissue arrangements (low power plans), (ii) labelled drawings to show the details of cell structure in the various tissues.

The following should be attempted—

Monocot. and dicot. primary stems and roots.
Dicot. dorsi-ventral leaf and petiole and monocot. isobilateral leaf.
Monocot. and dicot. stems and roots to show secondary growth.
Xerophyte and hydrophyte stems and leaves.

Permanent preparations of root and stem apical meristems, lateral root development, lateral bud structure and leaf abscission should be examined and drawn.

Floral Morphology and Anatomy

1. What are the characteristic reproductive organs of the angiosperm?

2. At which points on the body may these arise?

3. What is an inflorescence?

4. Distinguish between a peduncle and a pedicel, and between a bract and a bracteole.

5. By the use of suitable diagrams show the essential difference between racemose (indefinite) and cymose (definite) inflorescences.

6. Name and make diagrams of five forms of racemose inflorescences. Give at least one example of each.

7. Repeat for three forms of cymose inflorescences.

8. What is a verticillaster and a panicle? Give examples.

9. What other forms of inflorescences do you know?

10. Make a large, clearly labelled diagram to represent the arrangement of parts in a generalized flower.

11. Make a drawing of a longitudinal section through a named specimen which shows this arrangement.

12. What are the essential parts of the flower?

13. Summarize the main categories of characteristics by which flowers may vary from one another.

14. Distinguish between spiral and whorled and acyclic and cyclic arrangements of parts. Give examples.

15. What are the commonest numbers of floral parts?

16. How do monocots. and dicots. generally differ in this respect?

17. What is meant by a trimerous flower? Give an example.

18. What is the common number of whorls of parts in a flower?

19. What is considered to be the primitive condition of insertion of parts on the receptacle?

20. Distinguish between polypetalous and gamopetalous; apocarpous and syncarpous; adelphous, syngenesious, epipetalous, gynandrous and free stamens; introrse and extrorse stamens.

21. Make suitable diagrams to represent the hypogynous, epigynous and perigynous conditions. Give examples of each.

22. Distinguish between a superior and an inferior ovary. Give examples of each.

23. Distinguish between actinomorphic and zygomorphic; radial and bilateral symmetry; regular and irregular. Give examples.

24. Distinguish between hermaprodite and unisexual flowers; monoecious and dioecious plants. Give examples.

25. In which order of succession do floral parts usually arise on the receptacle? Give the name to describe this succession and also its antonym.

26. In what ways may the fusion of parts be brought about?

27. How does the pedicel and receptacle anatomy compare with that of a stem? Sepal and petal, with leaf?

28. Make a large clearly labelled drawing to represent the structure of a ripe anther as seen in T.S.

29. Of what units is a gynaecium composed?

30. Make diagrams to represent at least three common ways in which these may be arranged.

31. What is a loculus? How is it formed?

32. Name at least five possible forms of ovule placentation. Give examples of each.

33. Name three forms of ovule and make a diagram of the structure of each.

34. Distinguish between simple and compound styles. Give examples.

35. What is a nectary? State where it may commonly be found in flowers, quoting examples.

36. What happens to flower parts when they have performed their functions?

PRACTICAL EXERCISES

1. Dissect and elucidate the morphology of as many flowers as possible.

The following examples have been selected as representative of the commoner angiosperm families and at least one from each group should be studied in detail.

Identification should be made by use of a flora.

The floral morphology of each specimen should be recorded in the form of a written description, labelled drawing of a half-flower or a longitudinal section of the flower, a floral diagram and the floral formula.

The distinguishing floral characteristics of the families named should be noted and learnt during the completion of the practical work.

RANUNCULACEAE: Buttercup, wood anemone, lesser celandine, monkshood, columbine, larkspur.

CRUCIFERAE: Wall flower, shepherd's purse, cabbage, hedge mustard.

ROSACEAE: Dog rose, apple, wood avens, cherry, strawberry.

LEGUMINOSAE: Sweet pea, broad bean, clover, vetch.

CARYOPHYLLACEAE: Stitchwort, ragged robin, red campion.

SALICACEAE: Willow, poplar.

SCROPHULARIACEAE: Foxglove, germander speedwell, figwort, mullein.

LABIATAE: Deadnettle, sage.

COMPOSITAE: Dandelion, daisy, sunflower, cornflower.

LILIACEAE: Lily of the valley, bluebell.

IRIDACEAE: Iris, crocus.

GRAMINEAE: Perennial rye grass, wheat, oat.

2. Examine and make suitable drawings of permanent preparations of a selection of young flowers as seen in transverse section. Note especially detailed structure of anthers and gynaecia.

Reproduction and Life Cycle

1. Summarize the sequence of events which culminate in fruit formation.

2. Make a series of labelled diagrams to show the formation of pollen grains.

3. Describe in outline the cell division which immediately precedes pollen grain formation.

4. What is the significance of this division (genetical)?

5. What are the roles of the tapetum and the fibrous layers?

6. To what structure is the stamen equivalent in (*a*) *Selaginella* and (*b*) a gymnosperm?

7. Draw a diagram to represent the structure of a mature pollen grain as seen in optical section.

8. To what is the pollen grain equivalent in (*a*) *Selaginella* and (*b*) a gymnosperm?

9. How far can it be considered to be a reduced structure?

10. Make a series of labelled diagrams to represent the formation and development of an ovule up to the fertilization stage.

11. What is the origin of the megaspore mother cell?

12. What form does the division of this take and what are its products?

13. Into what does one of these products develop?

14. To what is this equivalent in (*a*) *Selaginella* and (*b*) a gymnosperm?

15. To what is the whole ovule equivalent in (*a*) *Selaginella* and (*b*) a gymnosperm?

16. How does an angiosperm ovule differ in form and position from that of a gymnosperm?

17. Draw a large, clearly labelled diagram to represent the structure of an ovule as seen in L.S. just prior to fertilization.

18. State the roles of its main parts and their destiny subsequent to fertilization.

19. What is meant by pollination? By what agencies may it be achieved? Give named examples of three different pollination mechanisms which you have watched. Point out the adaptations.

20. Summarize the floral characteristics usually found associated with the different pollination mechanisms.

21. Define the terms cleistogamic, cross-pollination, self-pollination, protogynous, protandrous.

22. Summarize the adaptations of flowering plants conducive to cross-fertilization. What are the genetical advantages of this?

23. What is the fate of the male gametes at fertilization?

24. Make a large, clearly labelled diagram to illustrate how they are delivered to the female structure. By what name is this manner of gamete delivery described?

25. What factors are known to influence the development of a pollen grain?

26. What is the origin of (*a*) the embryo and (*b*) the endosperm in the ripening ovule?

27. Summarize the stages in the development of an ovule into a seed.

28. Draw and label a series of diagrams to illustrate the formation of the embryo in a named angiosperm dicot.

29. Of what main parts is the embryo composed?

30. Of what fundamental parts is a seed composed? Make a labelled drawing of a named example.

31. Summarize seed forms with named examples.

32. What is perisperm? Where might it be found?

33. Distinguish clearly between a fruit and a seed.

*34. Fully define a seed in terms of its evolutionary origin.

35. Draw the characteristic life cycle diagram of a flowering plant.

36. Indicate how angiosperms exhibit alternation of generations.

37. How does this compare with the same phenomenon in (*a*) a liverwort or moss, (*b*) a fern, (*c*) *Selaginella* and (*d*) a gymnosperm?

*38. Give an outline of the possible evolution of the seed habit.

PRACTICAL EXERCISES

1. Examine permanent preparations and make suitable drawings of stages in the development of pollen grains (looking especially for meiotic divisions of the pollen grain mother cells), the passage of pollen tubes through stigmatic tissue, the embryo sac in a young ovule and the early stages of embryo development.

2. Make hanging-drop cultures of pollen grains in suitable sugar solutions and study their germination.

3. Make squash preparations of ovules of *Capsella* of various ages (starting from very young), look for and draw any stages in the formation of the embryo.

Fruit Morphology and Fruit and Seed Dispersal

1. From what part of a flower does a fruit develop and what conditions are usually necessary to its reaching maturity?

2. Distinguish between a true fruit and a pseudocarp.

3. What are the main parts of a true fruit?

4. What two main factors may be considered to determine the mature form of any true fruit?

5. Compile a table of classification of fruits based on their morphological structure.

6. Make large, clearly labelled drawings of named examples of each of the kinds you include in the classification.

7. In which of the main categories may each of the following be placed—nut, caryopsis, samara, cypsela, lomentum, silicula, siliqua, carcerulus, regma, double samara, cremocarp? Give the distinguishing features of each, with named examples.

8. Distinguish between receptacular and inflorescent fruits, giving examples.

9. Give the correct botanical description of the fruits of the following plants—apple, buttercup, maize, marsh marigold, violet, raspberry, hogweed, sycamore, fig, banana, lemon, walnut, elder, mulberry, parsnip, shepherd's purse, dead nettle, sunflower, hazel, pea, clematis, strawberry, tomato, ash, delphinium, foxglove, wallflower, sage, geranium, hollyhock, vine, coconut, marrow, hawthorn, pineapple, sweet chestnut, iris, rose, dandelion, laburnum, beech, horse chestnut.

10. Indicate clearly how a fruit may be distinguished from a seed.

11. What are the advantages of efficient seed dispersal?

12. Summarize the main agencies by which fruits and seeds may be dispersed.

13. For each given, quote examples, with drawings, of fruits and seeds so dispersed, pointing out the ways in which they are adapted morphologically.

14. What special structural features associated with their dispersal are shown by the following—a pine seed, a dandelion fruit, a hornbeam nut, a willow-herb seed, a bur-marigold fruit, an agrimony fruit, a goose-grass fruit, a blackberry fruit, an elder fruit, a holly fruit, a wood sorrel seed, an alder seed?

15. In each case state which part of the fruit or seed shows modification to be associated with the dispersal mechanism.

16. By what mechanisms are the seeds of the following plants dispersed—violet, poppy, balsam, crane's bill, broom, squirting cucumber?

17. In what ways has man influenced plant distribution through acting as a fruit and seed dispersal agent?

PRACTICAL EXERCISES

1. Collect, examine by dissection and make suitable labelled drawings to show the morphology of as many fruits as possible.

In all cases look for and make note of any special adaptations which may be associated with dispersal of the fruits or the seeds they contain.

THE ANIMAL KINGDOM

The Protozoa

1. What are the distinguishing characteristics of this phylum?
2. What is the meaning of the word protozoa?

TYPE STUDIES

3. Write down the generic and specific names of the types you have studied.

In the case of each type named—

4. State the habitat in which each exists in its adult phase.

5. Describe the habitat in terms of permanence, temperature and light variation, supply of food and oxygen, body support and movement, salinity and pH.

6. Give the approximate size in microns.

7. Make large, clearly labelled drawings to show the adult phase.

8. State the functions of each body component or organelle you have illustrated.

9. Name, describe and where possible illustrate, the outer surface of the body.

10. List the structural and functional adaptations to environment, e.g. colour, size, shape, protection, feeding and locomotive structures, sensory and osmoregulatory structures.

11. State the mode of life, the food substances and how these are obtained and utilized.

12. Where and in what forms is food stored in the body?

13. How does defaecation take place?

14. State the source of oxygen and explain how it is obtained.

15. Illustrate and describe the method of locomotion.

16. State how growth is manifested.

17. Name the excretory products and state how they are eliminated.

18. How is osmoregulation carried out?

19. State any responses which are made to stimuli and state the biological advantage which may be gained from each response.

20. State what sensory structures are known and what stimuli they perceive.

21. Name any co-ordinating mechanisms which are known.

22. Comment on the level of behaviour.

23. Illustrate asexual reproduction by large, clearly labelled drawings.

24. Repeat for sexual reproduction.

25. Note any reproductive adaptations to environment, e.g. conjugation, motility of gametes, protection of zygote, motility of zygote, protection of developing stages and provision for their nutrition, survival under adverse conditions, dispersal mechanisms, etc.

26. Make a large clear diagram showing the life history, labelling all the important stages especially those of meiosis and fertilization, where these occur.

27. Note any features of economic importance to man, both in your types and in related species.

28. Note any features of significance in the lives of other organisms.
Additional for parasites in this phylum—

29. In each case, name the host or hosts.

30. State the structural modifications of the parasite compared with related free-living forms.

31. Repeat for physiological modifications.

32. Repeat for reproductive modifications.

33. What is the effect on the host?

34. What control measures (prevention and treatment) have proved to be of value?

35. Give large, clearly labelled drawings of the life-history.

36. State clearly how the parasite is transmitted from host to host.

37. Name at least three human diseases caused by parasites of this phylum.

GENERAL STUDIES

38. Place the genera you have named in their classes and distinguish briefly between those classes.

39. On what criteria are the protozoa classified? Criticize the classification.

40. Name at least one representative of as many of the following protozoan forms as you can—uninucleate, multinucleate (coenocytic), colonial, sessile, radially symmetrical, spherically symmetrical, bilaterally symmetrical, heterotrophic, autotrophic, saprozoic, symbiotic, parasitic.

41. Name at least one parasite from each class.

42. Give at least three examples of symbiosis involving protozoa.

43. State both members of the partnership and what each contributes to the well-being of the other.

44. What class of the protozoa is considered to be the most primitive. Why?

45. State the known degree of differentiation found in each type you have studied.

46. How far is division of labour exhibited by the protozoa?

47. How far can a protozoan be compared with (a) the whole body of a higher animal, (e.g. a mammal), (b) a single cell of a higher animal?

48. What modes of nutrition have you found in the protozoa?

49. Define binary fission, sporulation, syngamy, isogamy, anisogamy, sporozoite, merozoite (schizozoite), zygote, gamete, autogamy.

50. What is a cyst? Give examples of different kinds of cysts from the protozoa.

*51. Outline the life cycle of one protozoan which shows alternation of generations, i.e. haploid and diploid phases.

*52. What other groups of living organisms show definite relationship with the protozoa?

PRACTICAL EXERCISES

1. Examine living specimens, measuring the size with an eye-piece micrometer.

2. Draw carefully, showing any visible features. Label clearly and state the magnification of your drawing.

3. Try to observe the feeding process and movement (slowed down). Carry out simple experiments on behaviour (e.g. light, chemicals, touch, obstacles, food substances, etc.).

4. Culture protozoa in the laboratory and try to observe reproductive stages.

5. Examine prepared slides.

6. Where suitable, make permanent stained preparations.

The Porifera (Sponges)

GENERAL STUDIES

1. What is the meaning of the word "porifera"?

2. What are the distinguishing characteristics of this phylum?

3. Write down the generic and if possible the specific names of any sponges you have studied.

4. State the general characteristics of the environments in which they are found (*see* No. 5 p. 37).

5. Name the classes in the phylum and distinguish between them.

6. Make large, clearly labelled drawings to show the morphology, anatomy and histology of a simple sponge.

7. State the functions of each part you have labelled.

8. Outline the physiology of a sponge under the headings of—nutrition, respiration, translocation, locomotion, growth, excretion, sensitivity.

9. Name at least three types of skeletal elements found in sponges.

10. What are the three grades of complexity? Illustrate.

11. Outline the methods of asexual and sexual reproduction.

12. Give a large, clearly labelled drawing of a sponge larva and describe its metamorphosis.

13. Of what economic importance are sponges?

14. With what group of the protozoa are they related? Why?

15. Describe the degree of regeneration of which sponges are capable.

16. Give a diagram of the life-history of a sponge.

17. How is dispersal effected?

18. What are the advantages of (*a*) radial symmetry, (*b*) the sessile mode of life?

*19. What features distinguish sponges from colonial protozoa?

20. Comment on co-ordination in sponges.

PRACTICAL EXERCISES

1. Observe and draw as many types of sponges as you can.

2. Examine prepared slides of sections of sponges. Draw any of the types of cells you can distinguish and also types of spicules.

The Coelenterata

1. What is the meaning of the word "coelenterata"?

2. What are the distinguishing characteristics of the phylum?

TYPE STUDIES

3. Write down the generic and, if possible, the specific names of the types you have studied.

In the case of each type—

4. State the habitat in which the animal exists.

5. Describe the habitat in terms of permanence, light and temperature variation, supply of food and oxygen, body support and movement, salinity and pH.

6. Give the approximate size when full-grown.

7. Make large, clearly labelled drawings showing the morphology, anatomy and histology.

8. State the function of each system, organ, tissue, cell and cell component labelled.

9. What is the structure of the wall between the external and internal media?

10. List the structural and functional adaptations to environment.

11. State the mode of life, the food substances and how these are obtained and utilized.

12. How is translocation effected?

13. How does defaecation take place?

14. State the source of oxygen and explain how it is obtained.

15. What skeletal structures are present and what functions do they perform?

16. How is locomotion effected?

17. State how growth is manifested.

18. How is osmoregulation carried out?

19. Name the excretory products and state how they are eliminated.

20. Describe any responses to stimuli and state any biological advantages gained by the responses.

21. What sensory structures are present and what stimuli do they perceive?

22. How is co-ordination effected?

23. Comment on the level of behaviour.

24. Illustrate asexual reproduction (if any) by large clearly labelled drawings.

25. Repeat for sexual reproduction and development.

26. Note any reproductive adaptations to environment, e.g. position of gonads, motility of gametes, food for the embryo, protection of zygote and developing embryo, survival under adverse conditions, dispersal.

27. Make a large, clear diagram showing the life-history, labelling all the important stages.

28. Note any features of economic importance to man.

29. Note any features of significance in the lives of other organisms.

GENERAL STUDIES

30. Place the genera you have named in their classes and distinguish briefly between these classes.

31. Name at least one representative of each of the following forms— hydroid alone, medusoid alone, both hydroid and medusoid forms in the life cycle, sessile, colonial.

32. Give at least two examples of symbiosis involving coelenterates. State both members of the partnership and what each contributes to the well-being of the other.

33. State the degree of differentiation found in the types you have studied, i.e. organ systems, organs, tissues, cells.

34. How far is division of labour exhibited?

35. What is meant by polymorphism? Give at least two examples.

36. State clearly the meaning of diploblastic.

37. Write short illustrated notes on nematoblasts, protandry, planula, statocysts, metagenesis.

38. With careful, labelled drawings show how the hydroid and medusoid forms are essentially similar.

39. Comment on the degree of regeneration of which coelenterates are capable.

PRACTICAL EXERCISES

1. Examine and draw living specimens.

2. Observe feeling, movement and behaviour.

3. Examine and draw preserved specimens and prepared slides.

4. Make your own permanent slides from small specimens or parts of larger ones.

The Platyhelminthes

1. What is the meaning of the word "platyhelminthes"?

2. What are the distinguishing characteristics of this phylum?

TYPE STUDIES

Repeat 3–29 as for coelenterates, pp. 40–1.

30. For No. 7 make separate, clearly labelled drawings showing— morphology, alimentary, excretory, nervous and reproductive systems, and transverse sections to show anatomy and histology.

Additional for parasites in this phylum—

Repeat 29–37 as for protozoa, p. 38.

GENERAL STUDIES

38. Place the genera you have named in their classes and distinguish briefly between those classes.

39. Give an example of symbiosis in this phylum. State the members of the partnership and what each contributes to the well-being of the other.

40. State the degree of differentiation found in the types studied, i.e. systems, organs, tissues, cells.

41. How far is division of labour exhibited?

42. State clearly the meaning of triploblastic.

43. Write short illustrated notes on rhabdites, flame cells, ootype (shell gland), miracidium, sporocyst, redia, cercaria, scolex, hexacanth embryo, cysticercus, proglottis.

44. Comment on the degree of regeneration of which planarians are capable.

45. Define spherical, radial and bilateral symmetry. Give examples of each and point out the advantages and disadvantages of each.

46. Is a tapeworm one animal or a colony? Give arguments for both views.

47. How would you distinguish between (*a*) a T.S. of a turbellarian, (*b*) a trematode, (*c*) a cestode?

48. Repeat for external surfaces of these three types.

49. Of what survival value is the flattened shape of the animals in this phylum?

50. What great biological advances do platyhelminthes show when compared with coelenterates? Give at least five.

PRACTICAL EXERCISES

1. Examine and draw free-living flatworms.

2. Observe and describe the movement, feeding and behaviour.

3. Examine and draw preserved specimens of trematodes and cestodes.

4. Examine and draw larval stages.

5. Examine and draw at least one transverse section of a member of each class studied.

The Annelida (Annulata)

1. What is the meaning of the word "annelida (annulata)"?

2. What are the distinguishing characteristics of this phylum?

TYPE STUDIES

Repeat 3–29 as for coelenterates, pp. 40–1.

30. For No. 7 make separate, clearly labelled drawings showing morphology, alimentary, blood vascular, excretory, nervous and reproductive systems, and transverse sections to show anatomy and histology.

GENERAL STUDIES

31. Place the genera you have named in their classes and briefly distinguish between those classes.

32. Name at least one representative of each of the following forms— a marine polychaete, a terrestrial oligochaete, an aquatic oligochaete, a leech, sexually dimorphic, hermaphrodite.

33. State the degree of differentiation found in the types studied, i.e. systems, organs, tissues, cells.

34. How is division of labour exhibited?

35. State clearly the meaning of coelomate.

36. What is meant by metameric segmentation?

37. Write short illustrated notes on—clitellum, chaetae, spermothecae, peritoneum, typhlosole, chloragogenous cells, light cell of Hess, parapodia, trochophore (trochosphere).

38. Name at least four implications of the coelom.

39. Comment on the degree of regeneration of which earthworms are capable.

40. Give at least four great biological advances shown by annelids when compared with platyhelminthes.

PRACTICAL EXERCISES

1. Examine live specimens, noting feeding, movement and behaviour.

2. Examine and draw external features.

3. General dissection showing as much of the internal structure as possible.

4. Dissection and drawing of the following systems—alimentary, blood vascular, excretory, nervous, reproductive.

5. Examination and drawing of transverse sections.

6. Make permanent stained preparations of nephridia, portion of nerve cord, ovary, smear from vesicula seminalis (earthworm).

The Arthropoda

1. What is the meaning of the word "arthropoda"?

2. What are the distinguishing characteristics of the phylum?

TYPE STUDIES

Repeat 3–29 as for coelenterates, pp. 40–1.

30. For No. 7 make separate clearly labelled drawings to show the external morphology (especially appendages), the alimentary, respiratory, blood vascular, excretory, nervous and reproductive systems, and transverse sections to show anatomy and histology.

GENERAL STUDIES

31. Place the genera you have named in their classes and briefly distinguish between the classes of this phylum.

32. Name at least one representative of each of the following forms—a marine crustacean with larval form, a terrestrial crustacean, an ectoparasite, an insect vector, a social insect, a spider.

33. What is the condition of the coelom in arthropods?

34. State the degree of differentiation found, i.e. systems, organs, tissues, cells.

35. Illustrate the structure of the body wall of a named arthropod in T.S.

36. How is division of labour exhibited?

37. State clearly the nature of the haemocoel.

38. Comment on segmentation compared with that of annelids.

39. Write short illustrated notes on sclerites, articular (arthrodial) membrane, endophragmal skeleton, chitin, sensillae.

40. Illustrate the structure of the arthropod eye.

41. What is the importance of the exoskeleton in arthropods?

42. Distinguish between mosaic and superposition images.

43. Give clearly labelled drawings to show the structure of an arthropod joint.

44. Explain how the muscles move the joint.

45. Describe the process of ecdysis for one particular type.

46. List the different types of respiratory organs found in arthropods and state what features they have in common.

47. What endocrine glands are known in arthropods? State the functions of their hormones.

48. Give at least four biological advances shown by arthropods when compared with annelids.

GENERAL STUDIES

Crustacea

49. Make a list of the appendages in order from anterior to posterior.

50. State the functions of each appendage.

51. Point out any adaptive features in each appendage.

52. What is autotomy? Give one example.

53. Illustrate a statocyst and state its functions.

54. Comment on regeneration in crustaceans.

GENERAL STUDIES

Insecta

55. Illustrate the mouth parts of at least two different types, e.g. biting and sucking.

56. Name at least one of each of the following forms—carnivorous, herbivorous, blood-sucking, sap-sucking.

57. Give at least two examples of biological control achieved by using insects.

58. Draw and label a typical insect leg (pereiopod).

59. Give at least three examples of adaptive variants of the typical leg.

60. Show the structure of a typical insect wing and state how it functions.

61. Name at least two modifications of insect wings for purposes other than flying.

62. Name at least one example of insect mimicry.

63. Outline the organization of one group of social insects.

64. Comment on polymorphism.

65. Name at least two diseases transmitted to animals by insects.

66. Name one disease transmitted to plants.

67. Outline the life-history of one insect vector.

68. Outline the life-history of one insect ectoparasite.

*69. Distinguish between ametabolic, hemimetabolic, holometabolic insects and give one example of each.

70. What major features account for the success of insects?

PRACTICAL EXERCISES

1. Examine live specimens noting feeding, movement and behaviour.
2. Examine and draw external features.
3. Draw the separate appendages under magnification where necessary.
4. General dissection showing as much of the internal structure as possible.
5. Dissection and drawing of the following systems—alimentary, blood vascular, excretory, nervous, reproductive.
6. Make permanent stained preparations of salivary glands (insects), portion of nerve cord; unstained preparations of small appendages.
7. Where possible keep live specimens throughout the life cycle; observe and illustrate the stages.

The Chordata

1. What is the meaning of the word "chordata"?
2. What are the distinguishing characteristics of the phylum?

GENERAL STUDIES

3. What are the structure and functions of the notochord?
4. Where in the body is it situated?
5. What are visceral clefts? Comment on their functions in the chordates you have studied.
6. By means of labelled drawings compare the position of the heart and the main lines of circulation in chordates and non-chordates.
7. What is a true tail?
8. List at least six different uses of the tail in chordates.
9. Compare the position and nature of the central nervous system in chordates and non-chordates.
10. Name at least two essential differences between chordate and non-chordate limbs.
11. Name the two sub-phyla and distinguish briefly between them.
*12. State at least three lines of evidence which relate chordates with echinoderms.
*13. Draw up a simple table showing the probable relationships of the major animal phyla.
14. Give a fully-labelled drawing showing a generalized chordate animal.
15. Give two simple transverse sections showing the essential differences between a chordate and a non-chordate (annelid or arthropod).

The Acrania

1. What is the meaning of the word "acrania"?
2. What are the distinguishing characteristics of the sub-phylum?

GENERAL STUDIES

*3. Name the three classes and name one genus in each class.
*4. List the chordate characteristics possessed by these genera.
*5. Name three types of larvae of acraniates.
*6. What are the functions of the pharynx in the genera named?
7. Name at least one of each of the following forms of acraniates—pelagic, sessile, burrowing, colonial.

The Cephalochordata

1. What is the meaning of the word "cephalochordata"?
2. What are the characteristics of the class?

TYPE STUDIES

Repeat 3–29 as for coelenterates, pp. 40–1.

30. For No. 7 make separate clearly labelled drawings to show the external morphology, the alimentary, respiratory, blood vascular, excretory, nervous and reproductive systems and transverse sections to illustrate anatomy and histology.

Additional for *Amphioxus*.

31. Name at least three structures which show asymmetry.

32. Give clearly labelled drawings to show the differences between primary and secondary gill bars, both in side view and in transverse section.

33. Illustrate the endostyle in T.S. and state its functions.

34. Illustrate the structure of the atrium and comment on its functions.

35. What is the extent of the coelom in the adult?

36. List the features in *Amphioxus* which are (*a*) primitive, (*b*) specialized.

For Embryology see pp. 55–6.

PRACTICAL EXERCISES

1. Preserved whole specimens for external features. Some dissection can be done with a good dissecting lens or microscope.

2. Transverse sections through oral hood, pharynx, and intestine should be carefully examined and drawn.

The Craniata (Vertebrata)

1. What are the meanings of the words "craniata" and "vertebrata"?
2. What are the distinguishing characteristics of the sub-phylum?

TYPE STUDIES

(Applicable in turn to all vertebrates)

3. Write down the generic and specific names of the types you have studied.

In the case of each type—

4. State the habitat in which the animal exists.

5. Describe the habitat in terms of permanence, light and temperature variation, supply of food and oxygen, body support and movement, salinity and pH.

6. Give the approximate size when full-grown.

7. What is the nature of (*a*) the limbs, (*b*) the tail?

8. State the functions of (*a*) and (*b*).

9. Give a simple sketch illustrating the layers which compose the body wall.

10. List the types of glands in the skin and state what each secretes.

11. Name any skin derivatives other than glands. Illustrate.

12. What are the functions of the skin?

13. Of what materials does the skeleton consist?

14. Make large, carefully labelled drawings to show the structure of the following skeletal parts—the cranium and sense capsules, the visceral skeleton, the vertebrae and their different kinds, the pectoral and pelvic girdles, the front and hind limbs.

15. What is the type of jaw suspension?

16. What is the condition of the notochord in the adult? Illustrate.

17. Make a simple labelled drawing to show the extent and subdivision of the coelom.

18. State the mode of life, the food substances and how these are obtained.

19. Make a large carefully labelled drawing of the whole of the alimentary canal, i.e. from mouth to anus.

20. Illustrate the structure of the teeth and state their numbers and functions.

21. List the glands associated with the alimentary canal and state their functions.

22. Outline the physical and chemical processes of digestion.

23. How is the absorptive area of the gut increased? Illustrate.

24. State the products of digestion.

25. Outline their fate in the body.

26. What is the condition of the visceral clefts?

27. Make a large, carefully labelled drawing of the respiratory system.

28. Describe the method of breathing.

29. Show how the respiratory surface is increased.

30. Describe the structure of the blood.

31. Give carefully labelled drawings of the external and internal structure of the heart.

32. Compose a simple diagram showing the general course of the blood circulation.

33. Illustrate the condition of the arterial arches.

34. Give careful labelled drawings of the arterial and venous systems.

35. Illustrate the portal systems and state their functions.

36. Trace the course of the blood from the heart to these structures and back again—(a) the hind limbs, (b) the small intestine, (c) the head.

37. List the changes in composition of blood entering and leaving these organs—kidney, liver, lung (or gills).

38. Give a simple illustrated outline of the lymphatic system.

39. What is the condition of the myotomes?

40. How is locomotion effected?

41. How is growth manifested?

42. What is the average duration of life?

43. Give a careful, fully labelled drawing of the male urinogenital system.

44. What type of kidney is present?

45. What is the condition of the Wolffian duct, the Mullerian duct, any other urinogenital ducts?

46. List the excretory products and state how they are eliminated.

47. Trace the course of the urine from the kidneys to the exterior.

48. How is the function of osmoregulation carried out?

49. Trace the course of the sperm from the testes to the exterior.

50. Give a careful labelled drawing of the female urinogenital system.

51. What is the condition of the Wolffian duct, the Mullerian duct and any other urinogenital ducts?

52. Trace the course of the urine from the kidneys to the exterior.

53. Trace the course of the ova from the ovaries to the exterior.

54. What glands are associated with the male and female genital tracts? State their functions.

55. How is fertilization effected?

56. What is the period of development from fertilization to hatching (or birth)?

(For embryology, see pp. 55–6.)

57. Make careful, fully labelled drawings of the brain in dorsal and ventral views.

58. List the main parts of the brain with their functions.

*59. What are the principal commissures and correlation centres?

60. Name the cavities of the brain from anterior to posterior.

61. How does the structure of the brain show correlation with the mode of life?

62. Give a careful, fully labelled drawing of the side view showing the cranial nerves.

63. Draw up a table of the cranial nerves in four columns—number, name, kind, regions supplied.

64. Give a large, careful, fully labelled drawing of a transverse section of the spinal cord.

65. Comment on the level of behaviour.

66. List the various sensory structures and state their functions.

67. What individual features are there in the eyes, the ears, the olfactory and gustatory organs?

68. What is the condition of the autonomic system?

69. Name and locate the endocrine glands and tissues.

70. State the known functions of their hormones.

71. List the major features which show adaptation to environment.

72. Note any features of economic importance to man.

73. Note any features of significance in the lives of other organisms.

THE PISCES

1. What is the meaning of the word "pisces"?

2. What are the distinguishing characteristics of the class?

TYPE STUDIES

Repeat 4–73, pp. 46–8.

GENERAL STUDIES

74. Place each genus you have studied in its order and give the main characteristics of that order.

75. Illustrate the structure of placoid scales (dermal denticles).

76. Make a careful labelled drawing to show the lateral line system.

77. What are the functions of this system?

78. List the fins present and state their functions.

79. Correlate at least two types of tail with the mode of life.

80. How are ascent and descent in the water achieved?

81. Name at least one of each of the following forms—a demersal teleost, a pelagic teleost, a demersal selachian, a pelagic selachian, a freshwater teleost, a lungfish.

82. What do you understand by "the single blood circulation of fishes"?

83. Make clearly labelled drawings showing transverse sections through (a) the heart region, (b) through the kidney region.

84. What is a cloaca?

85. Why is the discovery of living coelacanths of importance?

86. Of what evolutionary importance were fishes?

87. Write short illustrated notes on siphon, claspers, spiracle, Leydig's glands, ductus cuvieri, pericardio-peritoneal canal, ampullae of Lorenzini.

PRACTICAL EXERCISES

1. Living specimens should, if possible, be examined in aquaria and their feeding, movement, breathing and behaviour described.

2. Drawings showing external features should be done.

3. The various parts of prepared skeletons should be examined and drawn.

4. The following dissections should be performed, illustrated and carefully labelled—general contents of abdominal cavity; alimentary canal; heart and afferent branchial arteries; pharynx, gills, efferent and epibranchial arteries; other main vessels of the arterial system; venous system; urinogenital system (both sexes), cranial nerves, brain (removed).

5. Thick transverse cuts should be studied, drawn and labelled.

6. Permanent slide preparations of skin and blood should be examined and drawn.

Students' own preparations should include a section of cartilage, a blood smear (if fresh specimen available), teased muscle and nerve, dermal denticles.

THE AMPHIBIA

1. What is the meaning of the word "amphibia"?

2. What are the distinguishing characteristics of the class?

TYPE STUDIES

Repeat 4–73, pp. 46–8.

GENERAL STUDIES

74. Place each genus you have studied in its order and give the main characteristics of that order.

75. What do you understand by (a) cartilage bones, (b) membrane (dermal) bones?

*76. List the bones of types (a) and (b) in the skull.

77. Make a large, clearly labelled drawing of the pectoral girdle and sternum

78. What is a pentadactyl limb? Illustrate the typical condition.

79. State the modification of the typical condition found in the front and hind limbs.

80. What is the meaning of homodont, polyphyodont?

81. Comment on the structure of the lung and on lung breathing.

82. What is the structure of the truncus arteriosus (= conus + ventral aorta)?

83. Name at least five important differences in the anatomy of the brain compared with that of a fish.

84. Name at least one example of each—a tailed amphibian, a tail-less amphibian and a limbless one.

85. Write short illustrated notes on—columella, urostyle, atlanto-occipital membrane, carotid labyrinth, bladder, glands of Schwammerdam, Gasserian ganglion, solar plexus, fat bodies.

PRACTICAL EXERCISES

1. The living animals should be observed and their feeding, locomotion, behaviour and habits noted.

2. External features should be drawn.

3. Drawings of all skeletal parts should be made.

4. The following dissections should be carried out then carefully drawn and labelled—General dissection, alimentary canal (mouth to anus), respiratory system, arterial and venous systems, urinogenital system, spinal and sympathetic nerves, removal of brain and spinal cord.

5. Slides of the following should be examined and drawn, V.S. skin, T.S. lung, V.S. stomach, T.S. small intestine, kidney, testis.

6. Students' own permanent preparations should include—smears of blood, and faeces, teased preparations of muscle and nerve, stretch preparations of bladder and mesentery, section of cartilage.

For Embryology *see* pp. 55–6.

7. Where possible, a nerve-muscle prep. should be demonstrated to show muscle twitch with sciatic nerve and gastrocnemius muscle.

THE REPTILIA

1. What is the meaning of the word "reptilia"?

2. What are the distinguishing characteristics of the class?

TYPE STUDIES

Repeat 4–73 pp. 46–8.

GENERAL STUDIES

74. Place each genus you have studied in its order and give the main characteristics of that order.

*75. What is a temporal fossa and what is its importance?

76. Make a large, clearly labelled drawing of the pectoral girdle and sternum.

77. What modifications of the typical pentadactyl limb are found?

78. Comment on the structure of the lung and on lung breathing.

79. Comment on the truncus arteriosus.

80. Distinguish between oviparity, viviparity, ovoviviparity.

81. Name the groups of extant reptiles.

82. State at least three corollaries of the evolution of a neck.

83. Give at least eight reasons for the early success of reptiles as terrestrial animals.

84. Name at least one of each of the following forms—a marine reptile, a flying reptile, a gliding reptile, a limbless reptile.

85. Write short illustrated notes on—odontoid peg, autotomy, palatal folds, Jacobson's organ, fat bodies, allantoic bladder, azygos vein.

86. Comment on adaptive radiation in reptiles.

PRACTICAL EXERCISES

As on p. 50 for amphibia.

THE AVES

1. What is the meaning of the word "aves"?

2. What are the distinguishing characteristics of the class?

TYPE STUDIES

Repeat 4–73 pp. 46–8.

GENERAL STUDIES

74. Place each genus you have studied in its order and give the main characteristics of that order.

75. What is the keel (carina) and what is its significance?

76. What are pterylae and apterylae?

77. Distinguish between procoelous, acoelous, amphicoelous, and heterocoelous vertebrae.

78. List at least five reptilian characteristics possessed by birds.

79. Give careful labelled drawings showing the structure of the types of feathers.

80. List the modifications of the typical pentadactyl limb in the front and hind limbs.

81. Distinguish between homoiothermy and poikilothermy.

82. What is a sesamoid bone? Name one.

83. Comment on the nature of the oesophagus and stomach in birds.

84. Describe the structure of the lungs and comment on the breathing process.

85. Write short illustrated notes on—pigeon's milk, uropygial gland, pygostyle, syrinx, jugular anastomosis, turbinals, pecten, cere.

86. How does a bird fly?

87. Four groups of animals have evolved true flight. Name them and show the structure of the wing in each case.

88. Name at least one of each of the following birds—flightless running, flightless swimming, diving, flighted swimming, carnivore, insectivore.

PRACTICAL EXERCISES

As on p. 50 for amphibia.

In addition, study of the types of feathers with microscopic examination of barbules on a flight feather.

THE MAMMALIA

1. What is the meaning of the word "mammal"?
2. What are the distinguishing characteristics of the class?

TYPE STUDIES

Repeat 4–73, pp. 46–8.

GENERAL STUDIES

74. Place each genus you have studied in its order and give the main characteristics of that order.

75. Name five types of hair.

76. How do mammals maintain a constant temperature?

77. Give at least five ways in which heat is lost from the body.

78. Write short illustrated notes on the diaphragm, teats, perineal glands, mediastinum, nictitating membrane, diastema.

79. By means of a careful diagram show the basic structure of the skull.

80. What are these—the cribriform plate, zygomatic arch, turbinal bones, sagittal crest, tympanic bulla, occipital condyles?

81. Name the auditory ossicles and outline their evolution.

82. Define heterodont and diphyodont.

83. Name the kinds of teeth, the sets, and give dental formulae for at least four mammals.

84. Illustrate the typical structure of a tooth. Briefly explain how it develops.

85. List at least eight new features found in the skull.

86. Where and what are—the intervertebral discs, the nucleus pulposus, the odontoid peg, vertebrarterial canals, metapophyses?

87. Make careful labelled drawings showing a rib with its dorsal and ventral articulation, and the structure of the sternum.

88. What are the pronate and supinate positions?

89. State the location, structure and functions of tonsils, epiglottis, Eustachian tubes, cardia, crypts of Lieberkuhn, Peyer's patches, receptaculum chyli.

90. Explain clearly the process of peristalsis.

91. List the parts of the respiratory passage from nostriis to air-sacs.

92. Describe the structures of the lungs and the method of breathing.

93. Draw up a table comparing inhaled and exhaled air in percentages of N_2, O_2, CO_2, H_2O: account for the differences.

94. What do you understand by diastole and systole?

95. Sketch a typical mammalian growth curve: label the main parts.

96. Make a careful labelled drawing of the eye in horizontal L.S.

97. By means of careful labelled drawings show the taste areas on the tongue, the sensory parts of the tongue, nose, eyes and skin.

98. How are these effected—adjustment of lens, adjustment of pupil, formation of images on the retina, movement of the eyeball, protection of the eyeball?

99. Name and explain briefly at least five common defects of the eye.

100. Give a large, careful, fully labelled drawing to show the structure of the ear.

101. List the sense organs of the ear and state their functions.

102. Write short illustrated notes on the location, structure and functions of—fovea centralis, lachrymal glands, Meibomian glands, tensor tympani muscle, Eustachian valves, ductus endolymphaticus, fenestra rotunda.

103. Tabulate the endocrine structures under these headings—name, location, hormones, functions.

104. Name at least five diseases due to hormone deficiency.

105. State the location and known functions of the corpora striata, the thalamus, hypothalamus, Pons Varolii, cerebro-spinal fluid, crura cerebri, corpora quadrigemina.

106. Give at least eight major reasons for the success of mammals.

107. Comment on adaptive radiation in mammals.

108. Name at least two animals in each of the following orders—Monotremata, Marsupialia, Insectivora, Rodentia, Lagomorpha, Carnivora, Artiodactyla, Perissodactyla, Chiroptera, Subungulata, Cetacea, Sirenia, Primates.

PRACTICAL EXERCISES

As on p. 50 for Amphibia.

In addition—general dissection of the thorax, dissection of the neck.

The Histology of Mammals

1. Define clearly tissue, organ and organ-system with at least three examples of each.

2. State the general characteristics of epithelial, connective, muscular and nervous tissues.

3. Give tabular classifications of epithelial tissues (seven types), connective tissues (seven types), muscular tissues (three types), nervous tissues (five types).

4. For each of these types of tissues complete 5–8 below.

5. Give careful labelled drawings of a few cells, indicating their size.

6. Give at least three examples.

7. State the functions of each type.

8. Point out any adaptive features.

SPECIAL STUDIES

9. Name the two types of bones and state how each is formed.

10. State at least seven important substances to be found in blood plasma.

11. Name at least three ways in which blood differs from other connective tissues.

12. How does blood coagulate (clot)?

13. What is lymph?

14. Explain the basis of human blood groups.

15. Indicate the important uses of serology.

16. Write short illustrated notes on spindle organs, Pacinian corpuscles, Ruffini's end organs, bulbs of Krause, Meissner's corpuscles, end plates, synaptic knobs, Purkinje tissue.

PRACTICAL EXERCISES

1. Microscope preparations of all the main types of tissue should be examined and drawn carefully. Students should practise the following types of permanent preparation—smear, teased, stretched, sectioned.

Mammalian Glands

1. Define clearly gland cell, glandular tissue, glandular organ. Give at least three examples of each.

2. Distinguish, with examples, between exocrine, endocrine and holocrine glands.

3. Classify glands anatomically into five types and give three examples of each.

4. Indicate with examples what you understand by neurosecretion. For each of the glandular structures listed below—

5. State the mode of secretion, i.e. endocrine, exocrine, holocrine, the anatomical type, and the location.

6. Give a careful labelled drawing of the histology.

7. Name the substances secreted and state their functions.

8. How is secretion stimulated?

Types

Salivary, sebaceous, Harderian, pituitary, hypothalamus, prostate, mucous, sudoriparous, lachrymal, pineal, nerve endings, Cowper's, stomach (exocrine and endocrine), Meibomian, thyroid, mammary, lymphatic, Ebner's, Brunner's, ceruminous, parathyroids, thymus, liver, kidney (exocrine and endocrine), crypts of Lieberkuhn, perineal, supra-renal, duodenum (endocrine), testis, interstitial cells, ovary stroma, corpus luteum, pancreas (exocrine and endocrine).

PRACTICAL EXERCISES

1. The anatomy and histology of the main types of glands should be observed and drawn.

Mammalian Organs

For each organ in the following list—

1. State the location, colour, size and shape.

2. Give careful, fully labelled drawings of the anatomy (indicating also neighbouring structures), and of the histology.

3. Indicate the blood supply (both arterial and venous) and the innervation.

4. What are the functions?

5. State briefly (*a*) the embryonic origin, (*b*) the evolutionary origin (if known).

List of Organs

A muscle, a long bone, a nerve, a tooth, the tongue, stomach, small intestine, liver, pancreas, larynx, trachea, lung, kidney, bladder, penis, testis, ovary, brain, skin, eye, ear, nose, diaphragm, heart, artery, vein, spinal cord.

PRACTICAL EXERCISES

1. The anatomy and histology of the main organs should be observed and drawn.

Chordate Embryology

1. What are gametes, gametogenesis, genesis, spermatogenesis?

2. Name the phases of gametogenesis, pointing out clearly where meiosis occurs.

3. Give, in diagram form, fully labelled, an outline of oogenesis and spermatogenesis in the general case.

4. Classify ova with respect to (*a*) amount of yolk present, (*b*) the degree of cytoplasmic differentiation (presumptive areas).

5. Explain, with respect to cleavage, the terms, holoblastic, meroblastic.

6. What are micromeres and megameres? Illustrate with one example.

7. Distinguish, with examples, between birth and hatching.

8. Define clearly, with at least two examples of each—embryo, free embryo, larva, foetus.

9. Define clearly, gastrulation, neurulation, organogeny.

10. What are a blastula, a zygote, a germinal vesicle?

11. Distinguish with examples between primary and secondary egg membranes.

12. Name the four body cavities in order of their formation and state the functions of each.

13. Define, with respect to gastrulation, the terms, invagination, epiboly, convergence, divergence.

For each type studied—

14. Give large carefully labelled drawings of the gametes, stating their sizes.

15. Give an outline description of mating behaviour.

16. How is fertilization effected?

17. List the effects of fertilization.

18. Illustrate the fertilized egg—zygote.

19. Name the type of cleavage and illustrate the first four divisions.

20. What effects does yolk have on cleavage?

21. Draw carefully an orientated V.S. of the blastula.

22. With side and dorsal views map out the presumptive areas.

23. Name the processes by which gastrulation takes place.

24. What is the first stage at which chordate identification is possible?

25. What is the first sign of metameric segmentation?

26. Illustrate gastrulation by four or five careful drawings labelled and orientated.

27. Give careful labelled orientated drawings of a completed gastrula in T.S., sagittal V.S. and horizontal L.S.

28. State the origin of the mesoderm and coelom.

29. Outline the process of neurulation and give careful diagrams of T.S. and sagittal V.S. of a completed neurula.

30. Illustrate simply the differentiation of the mesoderm.

31. Give an approximate timetable from fertilization to—end of the first cleavage, completed blastula, completed gastrula, completed neurula.

Additional for *Amphioxus*—
32. How does the formation of the coelom differ from that found in vertebrates?
33. How does neurulation differ from that found in vertebrates?
34. What are the first signs of asymmetry?
35. By means of drawings of successive transverse sections show the method of formation of (*a*) the atrium, (*b*) the pharyngeal bars.
36. Describe briefly the formation of the gill slits and the endostyle.
Additional for *Rana*—
37. What are the functions of the albumin?
38. What is the yolk-plug?
39. By means of careful, labelled drawings of side and ventral views of the following stages, show the external changes visible in development—

       (*a*) 7 days old—about 6 mm long;
       (*b*) the tadpole just before hatching;
       (*c*) fully developed external gills;
       (*d*) fully developed internal gills;
       (*e*) front and hind limbs visible;
       (*f*) the young frog—still with tail.

40. Summarize the changes which take place during metamorphosis with respect to the skin, the skeleton, the teeth, the gut, the fins, the tail.
41. Give three or four careful, labelled drawings to show the progressive changes in the arterial arches.
42. Write short illustrated notes on cement gland, operculum, stomodaeum, proctodaeum.
(For organogeny, *see* 52 below.)
Additional for *Gallus*—
43. Distinguish clearly between blastodisc and blastoderm.
44. Write short illustrated notes on the area pellucida, area opaca, and area vasculosa.
45. How can you orientate a fertilized egg with regard to the future axes of the embryo?
46. Describe carefully the formation, functions and significance of the primitive streak, knot, groove and pit.
47. What are extra-embryonic structures? Give four major reasons for their importance.
48. Describe the state of development at the time of laying.
49. How can the approximate age of an embryo be determined in a transparency preparation?
50. Summarize briefly the state of development at the end of the first, second and third days after laying.
51. Illustrate each of these by three careful labelled drawings—(*a*) a transparency, (*b*) T.S. through eyes, (*c*) T.S. through the heart.
52. By means of simple clear drawings outline the development of the blood vascular system (including the heart), the alimentary canal and its derivatives, the kidneys, the central nervous system, the eyes and ears, the vertebrae.

53. Name the embryonic membranes and state their functions.

54. By means of three clearly labelled drawings show the development of these membranes at early, middle and final stages.

55. State the kind of germ layer inside and outside each membrane.

56. What are the functions and final fate of the yolk and the albumen?

57. State how the chick hatches and what happens to the embryonic membranes.

58. Explain clearly what is meant by "cleidoic" development.

59. Write short illustrated notes on cranial flexure, anterior and posterior intestinal portal, somatopleur, splanchnopleur, umbilical stalk, egg-tooth.

Additional for mammal.

60. Make a careful labelled drawing of a V.S. of a blastocyst.

61. With what stage of lower vertebrates is the blastocyst homologous?

62. Describe the processes of temporary and permanent implantation.

63. Describe the stages in the formation of the embryonic membranes.

64. In what ways do the functions of these membranes differ from those of the chick?

65. What are the functions of the placenta?

66. Describe its formation and illustrate its final state.

67. List the stages in the birth of the mammal.

68. What is the fate of the embryonic membranes?

69. State the modes of nutrition throughout development.

70. Write short notes on gestation, parturition, trophoblast, umbilicus.

PRACTICAL EXERCISES

Amphioxus

1. Various stages should be examined in permanent preparations and drawn.

Frog

2. Mating and fertilization should be observed and the embryos reared as far as possible. From these, external features should be drawn.

3. Prepared slides of stages up to hatching should be drawn in sequence.

Chick

4. Examination and drawing, in sequence, of prepared slides of whole mounts and sections.

Mammal

5. Embryos at various stages should be examined if available. At the least an embryo attached to the placenta should be seen and drawn.

Water Relationships

1. For what two main reasons, in broad terms, is water of vital importance to living things?

2. List the properties of water which may be considered to have some biological significance.

3. In each case state what the significance may be and quote any relevant physical and chemical data.

Plants

4. What may be considered to be one of the greatest threats to survival of land plants?

5. In what two main ways may land plants be structurally adapted to survive this hazard?

6. Failing such structural adaptations give some examples of plants which still manage to survive.

7. How far can such plants be considered to be land-adapted?

8. Where is the main source of water supply for a land plant?

9. What features of this source contribute to its efficiency as a "reservoir" for plants?

10. What kinds of specialized water-absorbing organs may be developed by land plants? Which is the most advanced of these absorbing systems?

11. Note the morphological and anatomical features of this which contribute to its efficiency.

12. What forces must be overcome before this absorbing system can take water from its surroundings?

13. How may forces strong enough be developed in the plant cells?

14. Outline a method by which you could demonstrate that some plants are capable of exerting a water-absorbing force.

15. Define the term transpiration. How is the phenomenon related to evaporation?

16. Why may it not be regarded as a physiological function of the plant?

17. Why do many of the lower land plants die when exposed to drying conditions?

18. Why does this not apply to higher plants? What means of protection have they evolved against desiccation?

19. What contributes to the partial insufficiency of this protection?

20. What compromise between perfect protection against desiccation and efficient aeration of the tissues has been evolved?

21. Through which organs of a land plant is most water lost?

22. What features of their structure contribute to this?

23. Outline three ways in which this water loss may be demonstrated or measured. In the case of the latter specify the units in which the water loss would be expressed.

24. Give some estimates of quantities of water which may be lost in specified cases.

25. About what proportion of the water lost by a leaf passes through the stomata? By what name is other water transpired from the leaf usually known?

26. Make large, clearly labelled drawings to illustrate the structure of a stomate of a named plant.

27. What property of the stomate is associated with its capacity to restrict the passage of water vapour from the inside to the outside of the leaf?

28. Give a brief account of the mechanics of the movement of (*at least three forms of) stomatal guard cells.

29. Give a physiological explanation of these movements.

30. Would your explanation fit all known facts concerning stomatal movement? If not, give some possible exceptions.

31. Give at least six external factors which may influence transpiration rate.

32. In each case say what the influence is likely to be, quoting the relevant data to support your assumption.

33. Summarize the structural features of plants which may influence the transpiration rate. Quote specific examples and indicate by labelled drawings the nature of these features.

34. By what name do we describe plants which can live in the driest of habitats?

35. Give an outline of the conditions prevailing in at least two such habitats.

36. Define the term "transpiration stream."

37. Which are considered to be the water-conducting tissues of the plant?

38. Relate the structure of the cells of these tissues to their functions.

39. With reference to the parts played by epidermal and root hair cells, cortex, endodermis and pericycle, explain how water may be "actively" conveyed across root tissues.

40. How could you demonstrate that such a pressure may exist in a root and what is it called?

41. How far might the existence of this pressure account for the movement of water upwards through the plant?

42. What other forces may be considered to contribute to this movement?

43. Which of these appears to be of greatest significance in that it can be applied to all cases?

44. What structural and physical conditions of tissues and the moving water must be fulfilled if this explanation of the movement of water to the top of a tall tree is to be accepted?

45. How far do the observed conditions meet requirements?

46. Construct a large, clearly labelled diagram to represent the path taken by water in moving through a plant and annotate it to indicate the forces involved in maintaining the movement from point to point.

47. What may be considered as the force ultimately responsible for the movement?

48. What is a hydathode?

49. Explain the process of guttation. Give named examples of plants in which it is known to occur.

50. What beneficial effects have sometimes been ascribed to transpiration?

51. How far can transpiration be regarded as a necessary physiological function of the plant?

PRACTICAL EXERCISES

1. Measure the water content of various plant parts.
2. Measure the water-holding capacity of various soils.
3. Measure the "permanent wilting point" of a soil.
4. Compare the stomatal and cuticular transpiration from leaves.

5. Demonstrate the water-loss from a twig using a potometer, compare rates of loss under different external evaporating conditions and compare these rates with evaporation from an atmometer under similar conditions. Convert potometer readings into measured water losses.

6. Compare rates of water-loss from the two surfaces of a leaf using cobalt chloride paper and a standard colour paper.

7. Investigate the effects of external conditions on stomatal aperture using a porometer.

8. Measure the root pressure exerted by a potted plant.

9. Demonstrate the "lifting power" of transpiring leaves.

10. Demonstrate guttation.

11. Make epidermal strips and cut sections of leaves, examine and make drawings to show the detailed structure of stomata.

12. Examine cleared *Fuchsia* leaves and draw hydathodes.

13. Using a filter pump compare the rate of flow of water through equivalent cross-sectional areas of the wood of different trees. Correlate the results with the structure of the xylem elements.

Animals

1. Define carefully isotonicity, hypotonicity, hypertonicity.

2. How is tonicity of body fluids maintained in marine coelenterates, annelids, crustaceans, elasmobranch and teleost fishes?

3. What are the special osmotic conditions in the frog?

4. State at least three factors which reduce water loss in each of these groups of terrestrial animals—insects, reptiles, birds and mammals.

5. Explain clearly how these animals compensate for water loss from their bodies—rabbit, lion, desert rat.

Co-ordination

1. Explain what is meant by co-ordination as it is applied to the bodies of living things.

2. What are the primary functions of any plant or animal co-ordinating mechanism?

3. By what three main ways may co-ordination be effected by living things?

Plants

4. Which co-ordination mechanism is applicable to plants?

5. Give any reasons why plants appear to be less complicated than animals in this respect.

6. What is a plant hormone?

7. Distinguish between an auxin, a gibberellin and a cytokinin.

8. Name any parts of the plant in which they are known to be formed.

9. Name any parts of the plant where they are known to be effective and say what their co-ordinating roles may be.

10. Name one plant hormone which seems to be very widely distributed through the plant kingdom.

11. Name any others which are thought to exist.

12. Give the names of at least three synthetic compounds which are known to have comparable effects on plants.

13. Give five growth phenomena in which plant hormones are known to play some part.

14. In each case state briefly what that part may be.

15. Outline what is known of hormone movement through plant tissues.

16. Give at least three uses of plant hormones or related substances in horticulture and agriculture.

17. What concentration of hormones is usually effective?

PRACTICAL EXERCISES

1. Study the effects of β-indolyl acetic acid (in lanolin) when smeared on the cut surfaces of decapitated garden pea seedlings. Cut sections of the tips after several days, suitably stain and look for cell division.

2. Investigate the effect of the same substance in inducing growth curvatures in seedling stems.

3. Test the efficiency of hormone "weed killers."

4. Test the efficiency of "rooting powders" on plant cuttings.

Animals

1. State the three ways in which coordination may be effected in the higher animals.

(For endocrine glands and hormones *see* p. 54.)

2. Make a large careful drawing of any mammal in side view and mark the positions of endocrine glands and tissues. Label fully.

3. Summarize the main co-ordinating effects of hormones.

4. Distinguish clearly between organizers and evocators. Give examples.

5. What are pheromones? Can they be regarded as hormones?

6. Give examples of these functions in both invertebrates and vertebrates.

7. Define clearly, central nervous system, peripheral nervous system, sense organ.

8. Summarize the functions of the nervous system.

9. Give a clear fully labelled drawing showing the nervous connection between an exteroceptor and an effector.

10. Repeat for an enteroceptor and effector.

11. What do you understand by autonomic control?

12. What is a reflex action? Illustrate a named spinal reflex.

13. Write short illustrated notes on grey matter, white matter, dorsal ganglion, synapse, brain centres, commissures, correlation areas.

*14. Illustrate a simple apparatus for detecting a change in potential along a nerve fibre which is conducting an impulse.

*15. Give a simple graphical illustration of a nerve action potential and define spike, negative after-potential, positive after-potential.

16. What do you understand by "all or none" with respect to nervous conduction?

17. What are (*a*) the greatest frequency of nerve impulses, (*b*) the greatest speed of conduction?

18. What electrical changes take place in a nerve fibre as an action potential passes?

19. What do you understand by "facilitation of a nerve fibre"?

20. Illustrate the type of synapse between neurones (synaptic knobs).

21. What two conditions must be satisfied before an impulse will pass from one neurone to another?

22. What evidence is there of hormonal transmission at nerve-endings?

Translocation of Materials

1. What needs of the body are met by adequate transport systems?

Plants

2. What three main classes of materials are known to be actively transported by plants?

3. Outline the techniques which have been used to investigate the movement of mineral salts through the plant.

4. What has been discovered concerning the movement of these substances?

5. Outline a technique which has been used to investigate the movement of elaborated food materials through the plant.

6. How may the direction of movement of these substances differ from that of minerals?

7. What tissue is generally regarded as being responsible for the movement of sugars, amino acids, etc.?

8. What evidence is there for this?

9. What forces are presumed to be involved in the transport of materials through xylem?

10. What properties of xylem tissues are suited to the efficient working of such a transport mechanism?

11. What mechanisms have been suggested to account for movement of materials through phloem?

12. Enlarge on the likelihood of each of these as being of primary significance, giving reasons for and against.

13. What properties of phloem tissue may be correlated with its presumed method of functioning?

14. Summarize what is known of hormone movement through plants.

15. How do respiratory gases move through plants?

16. How does carbon dioxide reach photosynthesizing cells?

17. Relate the evolution of vascular tissue in plants with the "conquest of the land."

PRACTICAL EXERCISES

1. Investigate the effects of bark-ringing on growing twigs on trees and shrubs.

2. Study the position of development of roots on a bark-ringed willow twig cutting. Explain your results.

3. Study the effect on carbohydrate movement out of a leaf of severing the main veins near the petiole.

Animals

1. How can the movement of nutriment and oxygen in the body be explained in the cases of *Amoeba* and *Paramecium*?

2. Name at least two phyla in which external water is the translocating agent.

*3. Explain how food ingested by the choanocytes of a sponge reaches the outer cells—pinacocytes.

4. Illustrate and explain the system of water circulation in hydroid and medusoid coelenterate forms.

5. How are translocation of food, oxygen and excretory materials accomplished in Platyhelminthes?

*6. Comment on mechanisms for translocation in nematodes, rotifers and nemerteans.

7. Explain the mechanisms for translocation of food and oxygen in an aquatic arthropod (e.g. lobster, crayfish).

8. What translocation mechanisms account for the great activity of insects and of cephalopods?

*9. How are respiratory exchange and circulation of food achieved in echinoderms, e.g. starfish?

10. List the substances carried in the blood of vertebrates.

11. How is the circulation of the blood maintained?

12. What do you understand by neurogenic and myogenic hearts? Give examples.

*13. Give a careful labelled illustration of a mammalian heart showing the main bundles of Purkinje tissue.

*14. What are the sinu-auricular and auriculo-ventricular nodes?

15. How is the beating of a mammalian heart initiated and controlled?

16. What are the main differences between arteries and veins, structural and functional?

17. Give a careful, labelled drawing of the main features of a mammalian lymphatic system.

18. What are the functions of the lymphatic system?

19. Comment on the nature and functions of haemoglobin, *haemocyanin, *chlorocruorin, *haemerythrin.

20. In what groups of animals may these pigments be found?

Nutrition

1. For what main purposes are nutritional substances required by living things?

2. Wherein lies the fundamental difference between the nutritional processes of animals and green plants?

3. Are the nutritive requirements (in their final form) of plants and animals fundamentally different? Give reasons for your answer.

Plants

4. Name the ten so-called "essential" elements required by green plants.

5. Name at least five "trace" elements known to be required by green plants. Why are they referred to as "trace elements"?

6. Say what you can about the roles of all these elements in the structure and function of the plant.

7. What are the sources of these elements for the terrestrial green plant?

8. In what form and by what process is each of the elements you mention absorbed by the plant?

9. Briefly describe a way of showing that a particular element is essential to healthy growth of the plant.

10. If green plants are constantly absorbing these substances and incorporating them into protoplasm and other substances, why, under natural conditions, is there a balance between the amounts "bound" in living things and the amounts still available in their environments?

11. Construct a large, annotated diagram to illustrate the circulation of nitrogen in nature.

12. Clearly distinguish the three main sequences of events in this cycle and show how they influence one another.

13. Name the agencies by which nitrogenous material may be converted from one form to another and where possible give the chemical reactions.

14. Give two ways in which man influences the distribution of nitrogen.

15. Construct large, annotated diagrams to illustrate the circulation of (a) carbon dioxide, (b) sulphur and (c) water, in nature.

16. What is meant by carbon assimilation?

17. What is meant by photosynthesis?

18. Give at least five factors external to the plant which are known to influence the photosynthetic process.

19. Give at least two factors internal to the plant which also affect the process.

20. In all cases say what the influence may be, quoting relevant data.

21. Give three basically different ways of measuring the rate of photosynthesis. In each case state the units of measurement.

22. Summarize the evidence which indicates that there are both light and dark reactions or phases in photosynthesis.

23. What are the essential features of the light reaction? Give your answer in terms of ATP formation and the generation of a reducing agent.

24. What parts do light and chlorophyll play in this?

25. What light wavelengths are most effective?

26. What are the essential features of the dark reaction? How does it compare from an energy consideration with the respiratory reactions?

27. Construct an annotated chart to illustrate in outline the process of carbon assimilation by photosynthesis.

28. Write a simple chemical equation which represents the process overall.

29. What is the source of the oxygen evolved during photosynthesis and at what stage in the process is it formed?

30. In what way is the carbon incorporated into the synthesis?

31. What substance is among the first recognizable photosynthetic products?

32. Show briefly how this may be converted to carbohydrate.

33. Name at least two other main classes of compounds into which the carbon may be incorporated.

34. Say clearly why photosynthesis must not be regarded simply as a starch manufacturing process.

35. Construct a simple chart to show how the three main classes of organic compounds are metabolically interrelated.

36. Outline one way in which proteins may be formed from the products of photosynthesis.

37. Repeat for fats.

38. How does the photosynthetic activity of a green sulphur bacterium compare with that of a higher green plant?

39. How do the nutritional processes of fungi compare with those of (a) green plants and (b) animals?

40. Explain the term chemosynthesis and quote an example of its occurrence.

41. Explain the meanings of the following terms—limiting factor; hydroponics; photophosphorylation; oxidative phosphorylation; artificial fertilizer; compensation point; action spectrum; absorption spectrum; quantum of light energy; carbon dioxide acceptor molecule.

42. What parts do the following organisms play in the circulation of materials in nature—*Rhizobium leguminosarum; Azotobacter chroococcum; Clostridium pastorianum; Anaboena; Nitrosomonas; Nitrobacter; Thiobacillus denitrificans; Chromatium; Chlorobium; Thiobacillus thio-oxidans; Desulphovibrio*?

43. What two experimental techniques have contributed most to the elucidation of the biochemistry of carbon assimilation by green plants?

PRACTICAL EXERCISES

1. Investigate the effects of mineral deficiencies on the growth of plants using suitable culture solutions.

2. Demonstrate that starch is manufactured in leaves only in the presence of light, chlorophyll and carbon dioxide.

3. Measure the rate of photosynthesis in terms of oxygen evolved (bubbling aquatics), carbon dioxide uptake and increase in dry weight of tissues. Study the effects of temperature and light intensity on the rate of the process.

4. Make a fresh chlorophyll extract in acetone. Examine its absorption spectrum and separate the pigments by simple paper chromatography.

Animals

1. Name the six essential substances necessary in the diet of animals.

2. Indicate the chemical structure of these substances.

3. State the importance of each in the body.

4. Name three natural sources for each of these substances.

5. Tabulate the main vitamins under the heading of name, chemical nature, functions, sources, deficiency diseases.

6. Tabulate the mineral ions necessary in an animal's diet under the headings of name, source, functions, deficiency conditions.

7. What evidence is there that CO_2 can be fixed by animals?

8. Define the following terms—ingestion, digestion, absorption, assimilation, egestion.

9. Name at least five methods of ingesting small particles as food, with examples of animals which employ each.

10. Name at least three methods of dealing with large masses of food material, with examples of each.

11. What animals take in fluids? Explain the mechanism in three distinctly different cases.

12. Outline the physical processes of digestion in those animals you have studied.

13. Tabulate the chemical processes of digestion in a mammal under the headings—digestive juice, constituents, food acted on, pH, products of digestion.

14. Define extra-cellular and intra-cellular digestion with examples of each.

15. Comment on the presence of micro-organisms in the gut of animals as symbionts.

16. How are protein, fats, carbohydrates broken down in digestion?

17. State how the products are absorbed and their ultimate fate in the body.

18. List the various mechanisms you have encountered for increasing the absorptive area.

19. What are faeces? How are they eliminated in the animal types you have studied.

PRACTICAL EXERCISES

1. Experiments with digestive juices should be performed under controlled conditions. The action of proteases, amylases and lipases can be tested under varying conditions of pH and temperature. Enzymes and digestive juices (liquor pepticus, liquor pancreatini) can be obtained from biological dealers.

Respiration

1. From an energy viewpoint how does a living thing compare with a machine?

2. In what form may energy be stored in a living thing?

3. For what main purposes is energy required by a living thing?

4. Define the term respiration. Your answer must include a reference to ATP.

5. How is breathing related to the process?

6. What is a respiratory substrate?

7. Give examples together with their theoretical energy yields in kilo-joule per mole.

8. What is the meaning of oxidation?

9. Give three ways in which oxidation reactions may be brought about.

10. What is the meaning of reduction?

11. Write an equation to represent the oxidation of carbohydrate in air.

12. How does this compare with the oxidation of a carbohydrate in a living organism? State at least two fundamental differences.

13. Which groups of enzymes are concerned with oxidations and reductions?

14. Quoting examples, distinguish between the types of reactions catalysed by each of these enzyme groups.

15. Which group seems to play the major role in respiratory oxidations?

16. What are hydrogen acceptors?

17. Give two examples and show by simple equations how one of them may act in conjunction with an oxidizing enzyme to oxidize a substrate.

18. In the case given, show how the hydrogen acceptor may itself later be re-oxidized.

19. What part does atmospheric oxygen play in respiratory processes?

20. Distinguish between aerobic and anaerobic respiration.

21. Outline the main steps in the respiration of a carbohydrate substrate by both plants and animals (a) in the absence of oxygen and (b) in the presence of oxygen. (Full chemical details are not required.)

22. How far are these processes a continuation of one upon the other?

23. What are the end products in each case and how many ATP bonds are built per molecule of respired substrate?

24. What are the relative efficiencies of the processes?

25. In what form does the wasted energy of the substrate appear?

26. How does the organism restrict these losses?

27. How far may the respiratory process be considered to be the reversal of the photosynthetic process?

28. How closely is the fermentation of sugar by yeast related to the respiratory processes of higher organisms?

29. Distinguish clearly between oxidative phosphorylation and photo-phosphorylation.

30. Explain clearly what is meant by respiratory quotient and give the ranges of values for different substrates when respired under aerobic and anaerobic conditions.

31. What might you learn by measuring a respiratory quotient?

32. Explain clearly why diffusion is adequate for gaseous exchange in protozoa, sponges and coelenterates.

33. For each metazoan animal you have studied explain the methods of increasing the respiratory surface.

34. For each, outline the mechanism for avoiding stagnant conditions at the respiratory surface, i.e. breathing.

35. Illustrate the lung of a snail, the tracheae of an insect, the lung-book of a spider, the gills of a crustacean, the gills of a fish, the lungs of a frog and a mammal.

36. List the various devices for increasing the quantity of oxygen absorbed.

37. Explain clearly the mechanism for oxygen and carbon dioxide transport in mammals.

38. Why are germinating seeds most commonly used in demonstrating or measuring the respiratory activity of plants?

PRACTICAL EXERCISES

1. Demonstrate the gaseous exchanges which take place between plants and animals and their surroundings during aerobic respiration.

2. Repeat with plant material during anaerobic respiration.

3. Measure the rate of respiration of germinating seeds in terms of (a) CO_2 evolved per unit time, (b) loss in dry weight per unit time.

4. Repeat (*a*) for small animals.

5. Measure R.Q. for germinating seeds using a respirometer.

6. Repeat for small animals.

7. Determine the effect of change of temperature on rate of respiration for germinating seeds.

8. Demonstrate that heat is generated by germinating seeds.

9. Demonstrate the breakdown of stored food, e.g. starch grains in seeds during germination, by chemical tests and by direct observation. (Cut sections of dormant and germinating seeds and compare conditions of starch grains.)

Growth and Development

1. Distinguish between the growth of a population and the growth of an individual organism.

2. Draw a curve to represent the growth of a yeast or euglena population with time, under permanently perfect conditions.

3. Write the mathematical expression which fits this curve and say what your symbols mean.

4. Which law does the population growth rate obey?

5. How does this compare with the compound interest law?

6. Draw a curve to represent the growth of a yeast or euglena population with time under natural conditions.

7. How is this curve described and how does it vary from the curve drawn previously?

8. Explain this difference mathematically, i.e. say which factor is no longer remaining constant.

9. Give two reasons why a naturally occurring population would not grow according to the law described in 4.

10. How far do human population growth rates follow the pattern exhibited by yeast?

11. How would you define the real growth of an individual plant or animal?

12. Why cannot this be measured with exactness?

13. What two kinds of measurements are most commonly used to express the growth of an individual?

14. Give reasons why neither of these is perfectly satisfactory.

Plants

15. Outline methods of measuring the growth of roots, stems and leaves.

16. Outline a method of measuring the growth made by a whole plant throughout its life cycle.

17. Criticize your methods in all cases.

18. In what form would you express your experimental data?

19. What kinds of information about the growth of a plant could you expect to obtain?

20. Distinguish between growth and relative growth rate.

21. How do the values of these vary through the life of an annual plant?

22. Draw curves to represent the growth of (*a*) an annual plant and, (*b*) a perennial woody plant.

23. Distinguish between limited and unlimited growth, giving examples.

24. Interpret as far as you can the growth pattern of an annual plant (*a*) mathematically, (*b*) morphologically and, (*c*) physiologically.

25. Define the terms isometric growth, allometric growth, grand period of growth and senescence.

26. Give a summary of the ways in which growth is manifested by all the plants which you have studied giving details of where any growing points are located and how they are anatomically constructed.

27. Give at least five external factors known to influence growth in plants.

28. In each case show how variations in the factor may affect the process.

29. What is meant by etiolation? Fully describe an etiolated plant which you have seen.

30. What internal factors are known to influence plant growth?

31. How far can regeneration of tissues be regarded as a growth process in plants? Give three examples of regeneration of plant parts.

32. In what way may hormones influence the regeneration process?

33. Draw a series of labelled diagrams to represent the sequence of events during the healing of a wound by a woody stem.

34. What evidence is there that a wound hormone may play an active part in the healing process?

35. Give an example of continuous tissue replacement in a plant.

36. Describe the process in the example you quote.

37. What two main sets of factors influence the developmental pattern of a plant?

38. Name the five phases of development exhibited by an annual plant during its life.

39. What external conditions are necessary to successful germination and what internal stages of development must have been reached?

40. By what forces may water be taken up by a dry seed?

41. For what purposes is the water needed during the germination?

42. Why is oxygen required by the germinating seed?

43. What external sign may be given that the seed is using the oxygen for the purpose you mention?

44. How may variation in temperature affect the rate of germination and why should this be so?

45. Distinguish between light indifferent, light sensitive and light hard seeds. Give an example of each kind.

46. Distinguish between primary and secondary dormancy.

47. Give three possible causes of dormancy.

48. Explain what is meant by after-ripening.

49. Give one possible explanation of secondary dormancy.

50. Give a brief summary of the known facts relating to the viability of seeds.

51. Give a summary of the biochemical changes going on in a germinating seed.

52. Give an outline of the course of the vegetative development of a plant from its embryo condition to include both primary and secondary growth and showing the place and manner of initiation of all the main vegetative parts.

53. When is the reproductive phase considered to begin and end?

54. What three main factors are known to affect the onset of flowering?

55. What is meant by photoperiodism? State the known facts concerning it.

56. What is meant by vernalization? State the known facts concerning it.

57. In the general case upon what does the development of a fruit depend?

58. How far are hormones considered to be involved in this process?

59. What parts are played by hormones in any other of the developmental processes?

60. What is gibberellic acid and how does it affect the growth and development of plants?

PRACTICAL EXERCISES

1. Study the increase in numbers of a duckweed population.

2. Measure the length, area or volume (as appropriate) of plant organs at suitable time intervals. The auxanometer can be used very conveniently for a detailed study of growth in length of a stem or petiole, using short time intervals.

3. Measure growth of a plant by a dry weight method.

4. Express all results graphically, make an explanation of the form of the curve and analyse it for information concerning growth rates and relative growth rates at different times during the growth of the population, plant or plant organ.

5. Make observations on the healing of a wound in the trunk of a tree.

6. Keep an illustrated diary of the development of a suitable annual plant from seed germination to fruit formation.

7. Investigate the effects of external conditions on seed germination.

8. Germinate seeds in total darkness and make careful observations on the etiolated seedlings.

9. Study the effect of gibberellic acid on the growth of seedlings using an external spraying method.

Animals

1. Devise an experiment for observing growth in an animal by measuring length of the whole body or of a part of it.

2. How would you record your results graphically?

3. Devise an experiment for observing growth in an animal by (*a*) a volume method, (*b*) a dry weight method.

4. Give simple graphical representations of the growth pattern in a large crustacean, a small mammal, a human being.

5. Interpret these three growth patterns, mathematically and morphologically.

6. Name at least three types of tissue in the mammalian body where continuous multiplication is carried out.

7. What are the main external and internal factors which are known to affect animal growth?

8. Give a careful definition of regeneration.

9. Comment on the extent of regeneration in protozoa, sponges, coelenterates, turbellarians, earthworms, crustaceans, starfish, fishes, amphibia, reptiles, birds and mammals.

10. Describe the main stages in wound healing in mammals.

11. List the five main phases in the development of the vertebrate animal over the whole life span.

12. What are the main stages in development of the vertebrate embryo?

13. What part do hormones play in embryonic development?

14. Describe, with examples, at least three different methods of hatching or birth.

15. List the factors which affect the attainment of sexual maturity (*a*) in insects, (*b*) in amphibia, (*c*) in mammals.

16. How is metabolic balance achieved in adult mammals?

17. Define senescence and state some of its immediate causes.

18. What do you understand by gerontology? What is its importance?

PRACTICAL EXERCISES

1. Simple experiments on growth should be performed; graphs drawn and interpreted.

Locomotion

Animals

1. Name the three types of locomotion found in protozoa.

2. Illustrate, in named examples, the structures involved.

3. Explain each of these types of locomotion.

4. Give examples of the same kind of motion in higher animals.

5. Name at least three examples of protozoa which possess contractile power.

6. Illustrate the contractile tissue found in coelenterates.

7. What are the kinds of muscular tissue?

8. Differentiate between them histologically and functionally.

9. Distinguish, with examples, between phasic and holding muscles, flexors and extensors, protractors and retractors.

10. Illustrate the type of motor nerve ending found in voluntary muscle.

11. Give a careful labelled drawing of a simple apparatus for detecting electrical changes in contracting muscle.

12. Plotting millivolts against time illustrate the curve obtained by the previous experiment.

13. Explain this curve.

14. Outline the chemical changes which occur in contracting muscle.

15. Account for fatigue in a muscle.

16. Explain clearly, using one named example, the following types of movement—looping, creeping, swimming (sculling), swimming (rowing), walking (running, jumping), flying and gliding.

17. Name at least five animals from different groups which have evolved efficient gliding.

18. Write short illustrated notes on the following—tractellum, pulsellum, metachronal rhythm, tonus.

19. Explain clearly, and illustrate, how a muscle is attached to a bone.

20. Muscles are said to be arranged in opposing pairs. Can you quote any exceptions from the mammalian body?

21. What is the fastest recorded speed (km. p.h.) for these animals, each expert in its own class—cheetah, sailfish, swift, dragonfly?

Plants

1. Name three plants which make locomotive movements.

2. In each case say how the movement is effected.

3. Which parts of any *named* land plants make locomotive movements?

4. With what kind of environment are these movements associated?

Secretion and Storage

1. Define the term secretion and distinguish it from excretion.

2. Distinguish between intra- and extra-cellular secretion.

3. How would you differentiate between a storage substance and a permanent secretion?

4. Summarize the biological advantages of the ability to store energy and materials by plants and animals.

Plants

5. Name at least five different substances known to be secreted by plant cells.

6. In each case give the location and form of the secretory tissue and the presumed function of the secreted substance.

7. Explain the presence of oil in orange rind, resin in pine wood and "milk" in dandelion, poppy, spurge and rubber plant stems.

8. For what two main purposes is "food" stored in plant tissues?

9. Survey the structure of seeds in relation to their storage functions.

10. Survey, with named examples, the morphology and anatomy of roots, stems and leaves which are specially modified for storage purposes.

11. Name at least five different carbohydrate storage compounds found in plants.

12. In each case state the form in which it occurs, examples of plants in which it may be found and the precise location in the plant.

13. In what form are fatty substances stored in plants and in what tissues are they most commonly found? Give named examples.

14. In what form are nitrogenous substances stored in plants and in what tissues are they most commonly found? Give named examples.

PRACTICAL EXERCISES

1. Examine suitably stained or treated sections of fresh plant tissues and chemically test expressed plant juices to demonstrate the presence of all the commoner secretion and storage substances.

Animals

1. Give at least three named examples of animals which use secretions for the following purposes—trapping of food; assisting ingestion or swallowing; to facilitate gaseous exchange; to facilitate excretion; to facilitate fertilization; to construct the completed ovum; to feed the young; to provide skeletal materials; for defensive purposes, for offensive purposes, for construction of dwelling-places.

2. For each of the above cases, state the organs or tissues concerned with the secretion, their location, and the chemical nature of the substances secreted.

3. In what parts of the mammalian body are enzymes secreted?

*4. Name the enzymes associated with each part and state their functions.

5. What do you know of the secretory processes which take place within animal cells?

6. How do exocrine, endocrine and holocrine secretions differ?

7. Give at least three examples of nervous and of hormonal control of secretion.

8. Name at least three different animal groups in which $Ca\ CO_3$ is secreted as an exoskeletal structure.

9. Give examples of the secretion of silica, strontium sulphate, enamel, silk and wax.

10. For what purposes are food materials stored by animals?

11. What carbohydrates are stored in animal bodies?

12. State the locations where these are found.

13. Where is fat stored in insects, fish, amphibians, reptiles and mammals?

14. Why is fat, rather than carbohydrate, stored as an energy food in animal eggs?

PRACTICAL EXERCISES

1. Tests for storage substances in eggs, milk, liver, muscles, adipose tissue should be performed.

Excretion

Animals

1. How are excess water and excess mineral ions excreted by the animals you have studied?

2. Define clearly, with examples, ammonotelic, uricotelic, and ureotelic.

3. Where are excess amino-acids metabolized in mammals?

*4. Give a simple diagram showing the ornithine cycle of urea synthesis.

5. How is nitrogenous excretion carried out in earthworms, in insects, in crustaceans, in fishes, birds and mammals?

6. How is carbon dioxide excreted in the animal types you have studied?

7. Give three named examples of excretion by storage.

8. Name the excretory organs or methods of excretion in protozoa, sponges, coelenterates, flatworms, annelid worms, crustacea, insects, echinoderms, molluscs, cephalochordates, fishes, amphibia, reptiles, birds and mammals.

9. Give simple diagrammatic illustrations of these excretory structures.

10. Illustrate the processes which occur along the mammalian renal tubule.

11. Define the following terms clearly—ultrafiltration, resorption, osmoregulation.

PRACTICAL EXERCISES

1. The anatomy and histology of the various types of excretory organs should be examined and drawn.

2. A simple analysis of mammalian urine should be carried out.

Plants

1. Why do plants not exhibit excretory processes comparable with those of animals?

Sensitivity and Response

1. Define the terms sensitivity, stimulus, response and behaviour.

2. Summarize, under the separate headings external and internal, the stimuli to which (*a*) plants and (*b*) animals are known to make responses.

3. By what means, broadly, do plants and animals become aware of changing conditions?

4. In broad terms, how do plants and animals vary in the manner of detecting changes and in the manner of executing responses?

Plants

5. Give four ways in which sensitivity is manifested by plants.

6. Distinguish between paratonic and autonomic responses by plants, quoting examples.

7. What two main kinds of paratonic movements occur in plants?

8. For each of these briefly classify the movements and give examples of each kind.

9. Distinguish precisely between tactic, tropic and nastic movements.

10. Give at least three kinds of tactic responses according to stimulus and quote examples of plants exhibiting them.

11. In each case you mention, how is the response made apparent?

12. Give at least five kinds of tropic responses according to stimulus and quote examples of plants exhibiting them.

13. Distinguish between positive and negative response, orthotropic and plagiotropic response.

14. Of what biological advantage are tropic responses to gravity? Quote advantages gained by specific examples.

15. Outline two methods of demonstrating that a plant part is responding to gravity.

16. Give two ways of showing that a root tip is sensitive to gravity.

17. How may the region of response to gravity be located in a root?

18. What form does the reponse take? Explain it in terms of growth.

19. What part are auxins presumed to play in plant responses to gravity in (*a*) roots, (*b*) shoots?

20. Of what biological advantage are tropic responses to unidirectional light? Quote advantages gained by specific examples.

21. Outline a method of demonstrating that such responses are made by plant parts. Which plant organs are very convenient to use for this purpose?

22. Give one way of showing that a shoot tip is sensitive to unidirectional light.

23. How may the region of response to this stimulus be located in a shoot?

24. What form does the response take? Explain it in terms of growth.

25. What part are auxins presumed to play in responses to light stimuli by shoots?

26. Summarize the work of F. W. Went in this field.

27. Summarize what you know of the other tropic responses listed in 12 and indicate how you could demonstrate each of them.

28. Give at least five kinds of nastic responses according to stimulus and quote examples of plants known to exhibit them.

29. In each case indicate how the response can be demonstrated.

30. In each case say what form the response takes and explain it in terms of growth or other physiological phenomenon.

31. Outline the biological advantages of the nastic movements mentioned.

32. Give at least one example of an autonomic movement made by a plant stem.

33. Indicate how the movement may be demonstrated and quote examples of plants which may be used for the purpose.

34. What are a clinostat, an Avena-Einheit, a coleoptile, *Mimosa pudica*, a "sleep" movement, a pulvinus?

PRACTICAL EXERCISES

1. Study the tactic responses to light made by motile green algae.

2. Study the tropic responses made by roots, shoots, coleoptiles and tendrils. Include the use of a clinostat.

3. Study the nastic movements made by leaves and flowers.

4. Study the movements made by the stem of a climbing plant.

Animals

1. Name at least five different kinds of response exhibited by animals. Give examples.

2. Distinguish between exteroceptors, enteroceptors and proprioceptors.

3. Name the organs of chemoreception in the animals you have studied.

4. Give simple illustrations of the olfactory organs in a fish, an amphibian, a reptile, a bird and a mammal. What do you understand by the gustatory sense? Where are the organs concerned with this in fish, amphibians and mammals.

5. Illustrate the gustatory organs of a mammal.

6. Name the organs concerned with photoreception in each of the animals you have studied.

*7. Outline the chemical changes which take place in visual purple (rhodopsin) during exposure to light and on recovery. Give a brief account of the Young-Helmholtz theory of colour vision.

8. What conditions in human beings are deuteranopia, protanopia, tritanopia?

9. What are the functions of the visio-perceptive and visio-psychic areas of the brain?

10. Name at least three adaptations of the eyes in diurnal animals and in nocturnal animals.

11. What adaptations are found in the eyes of terrestrial as compared with aquatic vertebrates?

12. What is meant by binocular and monocular vision? Give examples.

13. What properties of sound waves are distinguished by the mammalian ear?

14. How are these perceived?

15. What other senses, apart from sound, are located in the mammalian ear?

16. What are neuromast organs? In what groups of animals are they found?

17. What stimuli are perceived by these organs?

18. Outline the importance of these neuromast organs to the animals possessing them.

19. In teleost fishes, how are pressure changes perceived, what is the reponse and how is it effected?

20. What effects on terrestrial animals are exhibited by (a) increased atmospheric pressure, (b) decreased atmospheric pressure?

21. What different branches of the tactile sense are found in mammals?

22. List the organs concerned with their perception.

23. What is the value, to man, of the sense of pain?

24. List the organs concerned with equilibrium perception in coelenterates, in crustaceans, in insects and in vertebrates.

25. What are chromatophores? Name at least six animals from different groups which possess them.

26. Distinguish between morphological and physiological colour changes.

27. Name at least four external factors which cause colour changes in animals.

28. Give examples of hormonal and of nervous control of chromatophores.

29. Distinguish, with examples, between inherited and learned behaviour.

30. What animals show uncoordinated responses to stimuli?

31. Define the following types of reflex action giving one example of each—kinesis, orthokinesis, klinokinesis.

32. What is a taxis? Define, with examples, chemotaxis, phototaxis, rheotaxis, thermotaxis.

33. Give a simple account of phobotaxes in protozoa.

34. What is "appetitive behaviour"? Give some examples and indicate how each may serve some homeostatic function.

35. Define conditioned reflex, selective learning and insight.

36. What do you understand by the term "intelligence"?

PRACTICAL EXERCISES

1. Simple experiments on responses to stimuli can be performed with most animals.

2. Observations of instinctive behaviour should be made.

3. Small mammals and other animals can be taught and the power of memory tested.

4. Record should be made of any cases of intelligent behaviour.

Reproduction

1. Distinguish precisely between the following, citing cases in which the terms are used correctly—sexual and asexual; hermaphrodite and unisexual; male and female; homothallic and heterothallic; monoecious and dioecious; cross-fertilization and self-fertilization.

2. Clearly define the following terms—gamete, zygote, spore, isogamy, heterogamy, merogamy, hologamy, sexual dimorphism, parthenogamy, autogamy, pseudogamy, parthenogenesis, paedogenesis, paedomorphosis, syngamy, fertilization, copulation, conjugation, neoteny, vegetative propagation.

3. Of what biological advantage is a rapid and efficient method of reproduction?

Plants

4. Summarize and place in sequence the events leading to the initiation of a new generation by the sexual process in plants.

5. How is sexual maturity recognized in flowering plants?

6. What external and internal conditions are known to influence the plant in reaching this state?

7. What is meant by a flowering-inducing hormone?

8. Survey the evidence which indicates the existence of such a substance.

9. Outline a hypothesis which has been advanced to account for the phenomenon of photoperiodism in short-day plants.

10. In what other ways besides an effect on flower formation may day-length influence some plants?

11. Account for the effect of vernalization on the winter rye.

12. How does day-length interact with temperature in this case?

13. In what organs are gametes produced in (*a*) flowering plants and (*b*) any other plants which you have studied? Draw labelled diagrams of any organs you mention.

14. What is the nature of the cell division immediately preceding the formation of pollen grains (microspores) and megaspores in the flowering plant?

15. Where does the corresponding division occur in life cycles of the other plants you have studied?

16. Draw a series of labelled diagrams to represent the behaviour of the chromosomes during this division.

17. Survey the occurrence of motility among plant gametes.

18. Draw and label a diagram to represent the fertilization process in flowering plants.

19. Draw similar diagrams to represent the same process in the other plants which you have studied.

20. Distinguish between zoidogamy and siphonogamy.

21. To which set of environmental conditions is each clearly an adaptation?

22. To what stimuli are motile gametes known to respond?

23. What physiological conditions are necessary to the successful germination of pollen grains?

24. What will be the nuclear condition and chromosome constitution of the cell resulting from the fusion of two gametes as compared with the parents?

25. Through what agency may the nucleus control the pattern of development of this cell and hence the final product?

26. Why may the final product of the fusion cell vary from the parents?

27. In what circumstances would it not do so?

28. Of what biological value is the introduction of variation among members of a species?

29. What is the advantage to a species of a sexual reproductive process?

30. Give any disadvantages of the process.

31. How are these overcome?

32. Broadly distinguish between two asexual processes in plants.

33. For each of these, note the form it takes in all the plants you have studied.

34. In what fundamental way does an asexual spore differ from a gamete?

35. Give the kinds of asexual spores formed by all the plants which you have studied.

36. Make a survey of all other kinds of asexual reproductive structures which you have encountered in plants.

37. Is asexual reproduction of any significance in introducing variation among members of a species?

38. Which reproductive process does a gardener employ when he wishes to maintain similarity in all successive generations? Give some examples of such propagation methods.

39. Which reproductive process does he employ when he wishes to breed new varieties? Give reasons.

Animals

1. Outline the general sequence of events culminating in fertilization.

2. What hormones influence the development of the gonads in vertebrates?

3. Outline the secondary sexual characteristics for both sexes in man, domestic fowls, frogs.

4. Under what influence are these characteristics developed?

5. What changes take place in human beings at the end of the reproductive phase of life?

6. List the sequences in oogenesis and spermatogenesis in mammals.

7. Give large, clearly labelled drawings of a mammalian sperm and ovum.

8. State the functions of all the parts labelled.

9. Outline the methods of liberation of the gametes in those animals you have studied.

10. How do the testes of mammals descend into the scrotal sacs?

11. Name at least three cases where sperm are stored by the female for comparatively long periods.

12. What is the general reaction of the ovum to sperm penetration?

13. Describe the nuclear details of fertilization in one named case.

14. What do you understand by reproductive rhythm, menstruation, ovulation, pregnancy?

15. Outline the physiological and morphological changes which occur throughout the oestrus cycle in mammals.

16. How are implantation, parturition and lactation initiated?

17. Outline with examples from insects, fish, birds and mammals the provisions made for the offspring.

18. Describe the care of the young in three cases from different groups of animals.

19. Give examples of binary fission, multiple fission (sporulation).

20. Name another kind of fission process and say how it differs from those just mentioned. Give an example.

21. Define fragmentation as a method of reproduction. In what animal groups does it occur?

22. What is budding (gemmation)? Outline the sequence of events in two different cases.

23. Comment on the lack of asexual reproduction in the higher animals.

24. Write short illustrated notes on scyphistoma, strobilation, neoteny, colonies.

25. Define parthenogenesis and give three examples of its occurrence.

26. Describe conditions under which artificial parthenogenesis has been induced.

27. Explain how sex is usually determined in animals.

28. What physiological factors influence sex after fertilization?

29. What are intersexes? State what factors may cause them.

30. Give examples of sex reversal, (*a*) as a normal part of the life cycle, (*b*) as an abnormal effect of environment, (*c*) produced artificially.

31. What is gynandromorphism? In what group of animals is it common?

32. Summarize the factors which influence the development of sex in mammals.

Special Modes of Life

GENERAL

1. Explain what is meant by "making a living" and "mode of life."

2. Distinguish between autotrophic and heterotrophic organisms, quoting examples.

3. What sources of energy are available to living things?

4. What forms of material (broadly) are available to living things?

5. In the general cases, which of the sources of energy and which of the forms of material are used by plants and which by animals?

6. Distinguish between the various autotrophic forms according to their energy sources. Quote specific examples.

7. Give examples of plants which exhibit the same mode of life as animals.

8. Distinguish briefly between the free-living condition, saprophytism and parasitism.

9. What is meant by each of the terms—holophytic, holozoic, saprozoic, acultative parasite, obligate parasite and partial parasite?

10. What is a symbiotic relationship?

11. What is commensalism?

Parasitism

12. Write down the names of all the parasites which you have studied.

13. Widen your concept of parasitism in the following terms—range of forms which may be found in such association; range of graded conditions in which it may occur; range of physical contact between host and parasite. Quote examples of parasites wherever possible.

14. What is the probable derivation of parasitic forms?

15. Suggest possibilities as to the ways in which parasitic relationships may have occurred.

16. Summarize the modifications from the free-living condition which may be exhibited by parasites under the following headings—the means of finding the correct host; the means of physical attachment to the host; the means of dispersal of offspring to new hosts.

17. Summarize the special characteristics of the parasites you have studied under the headings—structural; physiological; reproductive; life cycle.

18. Why does man attempt to control parasites?

19. What is the fundamental rule which must be obeyed if an attempt to control a parasite is to stand a chance of success?

20. Survey broadly the ways in which parasites may be controlled.

21. Within each main category of control measures give as many ways as possible of exercising the control.

22. For named examples of each of the following summarize all you know about their structure, physiology, reproduction, life cycle and possible control of—a fungal parasite; a higher plant parasite; a protozoan parasite; a flatworm parasite; an insect parasite.

23. Summarize the effects which parasites may have on their hosts.

24. What is meant by host tolerance?

Saprophytism

25. Write down the names of all the saprophytes which you have studied.

26. Widen your concept of saprophytism in the following terms—the range of saprophytic forms; the relationships between saprophytes and other living things.

27. In what way can saprophytism be associated with putrefaction and decay?

28. Summarize the part played by saprophytes in the circulation of materials in nature (e.g. nitrogen, sulphur, carbon).

29. Explain the possible origin of the saprophytic mode of life.

30. Summarize the special characteristics of saprophytes under the headings structural, physiological and reproductive.

31. How may some saprophytes bring about their own destruction?

32. Give two ways in which saprophytes influence man's economy.

33. In what ways can man control the saprophytes detrimental to his economy? Give examples.

34. How can man "improve on nature" by making use of saprophytes? Give examples.

35. Summarize all you know about two named saprophytes. Point out any special adaptations to mode of life.

Symbiosis

36. Write down the names of all the partners in the symbiotic relationships which you have encountered.

37. Widen your concept of symbiotic relationships in the following terms— the range of form of the partners; the degree of interdependence; the kinds of benefit to be derived from the partnership.

38. Explain the possible origin of symbiotic relationships.

39. Summarize all you know about at least two named symbiotic partnerships.

Commensalism

40. Distinguish clearly between commensalism and the symbiotic relationship.

41. Give examples of (*a*) nutritional, (*b*) positional commensals.

Insectivorous Plants

42. How do the nutritional processes of these plants compare with those of the normal green plant?

43. What advantages may insectivorous plants gain from their special nutritional characteristics?

44. What kind of special structural feature is exhibited by every insectivorous plant?

45. Summarize all you know about at least two named insectivorous plants, paying special attention to structural and physiological adaptations to mode of life.

Variation

1. Define the term variation.

2. How far is it a feature of living things?

3. Give an account of the degrees of variation to be encountered among living things, quoting examples.

4. Along what main lines have variation studies by biologists progressed?

5. To what uses have the results of any of the variation studies been put?

6. Distinguish between continuous and discontinuous variations, giving examples of each.

7. To what two main influences may variation among living things be traced?

8. What is meant by a biometrical study?

9. What is frequency distribution?

10. How may a frequency distribution curve be constructed?

11. What may the shape of such a curve tell the biometrician?

12. Draw the shape of a normal distribution curve.

13. Distinguish between mean and mode.

*14. Give two constants of a distribution curve which are of importance to the biometrician.

*15. Give the mathematical expression which fits a normal distribution curve.

*16. Which is the variable in this expression and what is it a measure of?

*17. What is standard deviation? Express it mathematically.

18. What is meant by correlation?

19. Give an example of how a knowledge of the degree of correlation between two conditions may be of value to the biologist.

20. Explain what is meant by the interaction of nature and nurture.

21. How is it possible to distinguish between the effects of these experimentally?

22. Summarize Johannson's work on pure lines. What did he deduce from his experiments?

PRACTICAL EXERCISES

1. Construct a frequency distribution curve, e.g. frequency distribution of length among a sample of beans.

2. From the data find the mean and mode values.

3. Complete the appropriate table and calculate the standard deviation.

4. Using suitable data, e.g. bean lengths and widths, plot a correlation dot diagram and estimate the degree of correlation between the two dimensions.

5. Complete the appropriate table and calculate the correlation co-efficient of the two dimensions.

Genetics

1. What aspect of biology is included in the study of genetics?

2. State clearly what Mendel set out to investigate.

3. What techniques did he use?

4. In the case of one pair of contrasting characters, what numerical distribution of the characters appeared in (a) the first filial generation (F1), (b) the second filial generation (F2)?

5. What was Mendel's explanation of these results? Illustrate the crosses made and the results obtained using symbols.

6. Why must you put a double "dose" of a factor in a body cell?

7. State Mendel's first law of inheritance.

8. Interpret this in your own words.

9. How can Mendel's F2 ratios above be compared with the ratios of head and tail pairings which result from tossing together two coins a large number of times?

10. In the case of two pairs of contrasting characters what numerical distribution of the combinations of characters appeared in (a) the F1, (b) the F2?

11. What was Mendel's explanation of these results? Illustrate the crosses made and the results obtained using symbols. (In doing this see that you obey Mendel's first law when arriving at the genetical constitution of the gametes.)

12. State Mendel's second law of inheritance.

13. Interpret this in your own words.

14. Combine Mendel's two laws and make one statement summarizing the extent of his discoveries.

15. In what fundamental way did Mendel's work differ from that of earlier plant and animal breeders?

16. Define the following terms using illustrative examples where possible—gene, allelomorphic characters, dominance, homozygous, heterozygous, hybrid, monohybrid ratio, dihybrid ratio, genotype, phenotype, reciprocal cross, back cross, test cross, clone, pure line, genome.

17. How many genotypes are there in the F2 of a cross involving two pairs of contrasting characters?

18. How many phenotypes and what are their proportions?

19. Name at least four cases of apparent deviation from the laws of Mendelian inheritance subsequently discovered by later geneticists.

20. Give an explanation of each of these and state what it tells us about the action and interaction of genes.

21. Diagrammatically represent the connection between the behaviour of the chromosomes during meiosis and the laws of Mendelian inheritance.

22. From a knowledge of the relative numbers of chromosomes and inheritable characteristics, what may be deduced concerning the distribution of genes on chromosomes (assuming that the genes are carried on the chromosomes)?

23. What name is used to describe this grouping of the genes?

24. Assuming it to occur, in what way is Mendel's second law invalid?

*25. Distinguish between coupling and repulsion.

*26. Using symbols to work out details of the crosses, contrast the expected distribution of characters in the F2 resulting from a cross between a heterozygote and a homozygote recessive (a test cross) for two characters (a) if there is no linkage, (b) if the two characters are totally linked.

*27. How does the practical experimental result of a cross involving two linked genes generally differ from the theoretical result?

*28. Explain this difference in terms of the behaviour of the chromatids during meiosis.

*29. What is meant by cross-over value?

*30. How can it be measured and what does it tell us about the distribution of genes on a chromosome?

*31. By what other means is it possible to construct a chromosome map?

32. Explain sex inheritance.

33. Quote at least one well-known case of sex-linked inheritance giving details of its transmission from one generation to the next.

34. Summarize all the evidence in favour of the hypothesis that genes are located on the chromosomes.

35. In what two main ways, from a knowledge of genetics, can we account for the origin of variation?

36. Which of these can be regarded as giving rise to true variation?

37. What name did de Vries give to the sudden changes in inheritable characters?

38. In what two main ways may these changes arise?

39. What is meant by polyploidy?

*40. Distinguish between auto- and allopolyploids, giving examples and stating how each may arise.

41. Summarize the structural changes which may occur in chromosomes.

42. Give an example of a gene mutation. How may one be artificially induced?

*43. What are plasmagenes?

*44. What is hybrid vigour?

*45. Outline the modern conception of chromosome structure (DNA). How does this fit the observed facts of cell division and account for the transmission of the genetic message from generation to generation?

PRACTICAL EXERCISES

1. Carry out simple breeding experiments with *Drosophila*.

2. Take part in a long-term project such as an investigation into the inheritance of flower colour.

3. Obtain a maize cob bearing grains of different colours. Count the grains in each colour and explain the proportions of coloured to colourless grains.

GENETICS QUESTIONS

1. Among the offspring of a pair of flies, 25 per cent have short wings and breed true for this character, the other 75 per cent having long wings. Of the latter, one-third breed true, the remainder being hybrids. What must have been the genetic constitution in terms of wing character of the two original parents? Explain your answer by diagrams and brief notes.

2. Two black female mice are crossed twice with the same brown male. In two successive litters, the first female produced a total of nine black and seven brown mice, while the second female produced seventeen black mice. What deductions can you make concerning the inheritance of black and brown coat colour in mice? What are the genotypes of all these mice?

3. In man, brown eye-colour (*B*) is dominant over blue (*b*). A blue-eyed man, with brown-eyed parents, marries a brown-eyed woman whose father was brown-eyed and mother blue-eyed. They have one child who is brown-eyed. What are the genotypes of all the individuals mentioned?

4. In tomatoes, the gene for red flesh (*R*) is dominant to the gene for yellow flesh (*r*). The gene for yellow skin (*Y*) is dominant to that for transparent skin (*y*). Given plants of the constitution *RRyy* and *rrYY*, how would you obtain pure-breeding tomatoes with red flesh and yellow skin? There is no linkage between flesh and skin colour.

5. A sand-hopper (*Gammarus*) has red eyes with a white surround. It is crossed with another sand-hopper with black eyes and no white surround. The offspring are of four kinds; (*a*) black eyes, white surround, (*b*) black eyes, no white surround, (*c*) red eyes, white surround, (*d*) red eyes, no white surround. Black is dominant to red, and white is dominant to non-white. Find the genotypes of the parents.

6. In the flowers of *Mirabilis jalapa*, red colour (*R*) is incompletely dominant over white (*r*). The heterozygotes are pink. If a red-flowered plant is crossed with a white one, what will be the flower colour of (*a*) the F_1, (*b*) the F_2, (*c*) the offspring of a cross of the F_1 with its red parent, (*d*) the offspring of a cross of the F_1 with its white parent?

7. Purple colour (*P*) is dominant to red (*p*) in sweet peas and round pollen (*R*) is dominant to short pollen (*r*). There is 87·5 per cent linkage between these two factors. Find the result of a backcross between the double recessive and individuals produced by mating homozygous purple round with homozygous red short.

8. What is the result in F_2 of crossing a female *Drosophila*, homozygous for white eyes, with a red-eyed male? Red is dominant to white; the character for eye-colour is sex-linked, and the male is the digametic sex.

9. and 10. In these two human pedigrees, the individuals shaded black possess the character mentioned. Squares represent males and circles females. Determine for each pedigree—(*a*) the genotype of each individual as far as possible, (*b*) the mode of inheritance of the character mentioned.

9. INHERITANCE OF MUSCLE ATROPHY

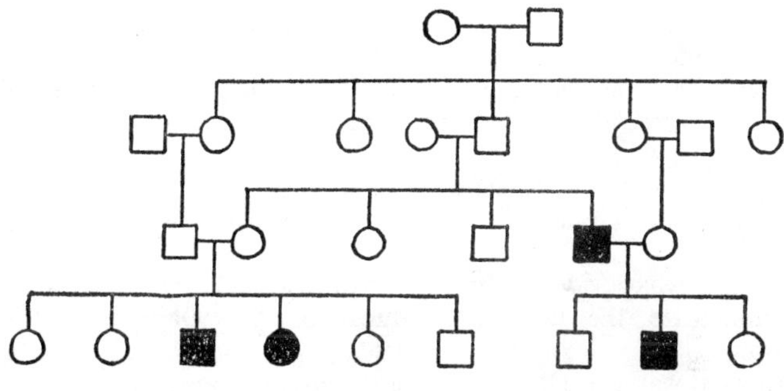

FIG. 1

10. INHERITANCE OF FEEBLE-MINDEDNESS

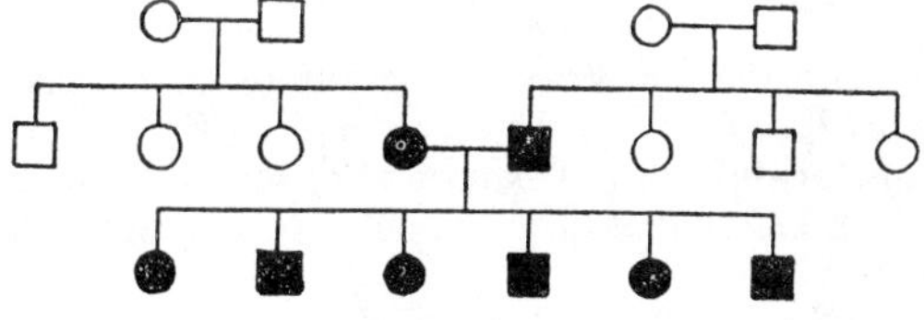

FIG. 2

Evolution

1. What is palaeontology?

2. Give the sequence of the major geological eras and their approximate durations.

3. Fill in the details of the geological periods within these eras if you can.

4. What is a fossil?

5. Summarize the processes leading to fossil formation.

6. Summarize the forms of fossils known to exist, giving examples where possible.

*7. What major groups of plants and animals are represented amongst the fossils found in the rock strata of each of the main geological periods?

*8. Construct simple charts to represent the time of origin and subsequent fate of the main groups of plants and the main groups of animals.

*9. To which present day groups of plants and animals are the following fossils related—

  Rhynia, Lepidodendron, Calamites, Etapteris, Lyginopteris, Cycadeoidea Homoioptera, Belinurus, Micraster, Hildoceras, Jamoytius, Ceratodus Eryops, Brontosaurus, Archaeopteryx, Oligokyphus, Proconsul.

10. What fundamental conception is embraced by the Theory of Organic Evolution?

11. Summarize, under at least four main categories, the evidence which supports the theory.

12. How may the study of (*a*) fossils, (*b*) homologous structures, (*c*) vestigial structures, (*d*) serology, contribute to the evidence for evolution? Quote examples of each of these.

13. What is meant by an acquired character?

14. Summarize Lamarck's theory of the inheritance of acquired characters.

15. Outline what evidence there is for or against Lamarckism.

16. Summarize the essential features of the Darwinian theory of evolution.

17. In what way was Darwin's theory similar to Lamarck's?

18. What was the weakest point in Darwin's argument?

19. What was Darwin's theory of "Pangenesis" and what bearing did it have on his theory of the Origin of Species?

20. Outline Weissmann's theory of the Continuity of the Germplasm.

21. In what way did this contradict previous ideas about the inheritance of variation?

22. What impact did de Vries' discovery of mutations have on the Darwinian theory?

23. How can Darwin's idea of natural selection be blended with modern conceptions of the inheritance of variation to form an acceptable theory of organic evolution?

24. Summarize the essential features of such a theory.

*25. Distinguish between major genes and polygenes.

Ecology

Note—

Since this is essentially a practical field subject there is no value in book learning of lists of factors and species of various habitats. The student can only revise those practical investigations in which he has taken part.

Man

1. Classify modern man in his species, genus, family, order, class and phylum.
2. Name other genera which are placed in the same family.
3. List at least five characteristics which distinguish man from the anthropoid apes.
4. By what criteria is man's evolutionary success judged?
5. List the features which have made man successful.
6. What do you consider to be man's greatest achievements?
7. What reasons can you give for man's phenomenal increase in numbers in historical times?
8. Name six inventions which you consider to have had far-reaching importance.
9. Give some account of man's treatment of the soil and point out the consequences.
10. What remedies are possible?
11. What large-scale effects have human beings had on natural vegetation?
12. Point out the consequences of these effects and state what remedies are possible.
13. Name at least three species of animals which have become extinct by human action.
14. Give some account of domestication of animals, and point out its effects, good and bad.
15. List at least three examples of biological control.
16. What is the present human population and what is its approximate rate of increase?
17. What problems does this raise? Make suggestions for their solution.
18. Do you believe there is a decline in average intelligence in the civilized countries? Substantiate your answer.
19. What evils may be attributed to biological ignorance?
20. What do you understand by disease? Classify diseases into nine groups.
21. Name some diseases whose incidence is (*a*) increasing, (*b*) decreasing.
22. Try to account for the increase or decrease in each case.
23. What dangers arise from increases in radioactivity?
24. On what grounds would you advocate a universal language?
25. State the obstacles.
26. On what grounds would you advocate universal biological education?
27. What would you include in the curriculum?
28. What steps would you advocate to increase human fitness for survival? Physical, mental and moral.

PART 3

THE EXAMINATION

To pass examinations well and with regularity calls for more than good luck. It is true that on occasion a paper may contain just those questions which the candidate has foreseen and made special preparation for, but it would be tempting providence to rely too heavily on "question spotting." There is only one road to unfailing success. *That is, by hard work, to come to know and understand all the relevant subject matter and then to display this comprehension to the best possible advantage.* It is to be hoped that by this time the student will, in his own honest estimation, have reached a high level of confidence in himself in the first matter. It now remains for him to become aware of how to acquit himself well in the second.

In order to do this there are certain facts which must be borne in mind and certain principles in the technique of examination-sitting which must be applied. But before going into detail of these, it might be of advantage to the candidate if he understood the reason for sitting an examination at all; if then he has serious doubts about his state of preparedness, he would be well advised to think again before sitting it.

A candidate sits an examination in order to prove to those who are in the best position to judge, the examiners, *that he has progressed along the path of study of his subject to a given stage from which he will be able to embark on further successful study at a higher grade.* The prospective candidate who knows clearly in his own mind that his preparation has been wholly inadequate would be well advised to reserve his effort for a future occasion and so save himself the disappointment which inevitably accompanies an examination failure.

The examiners signify their opinion of the state of the candidate's competence and progress by awarding a percentage mark. What is of extreme importance to the candidate is to know by what standards he is likely to be judged. The essential attributes of a first-class candidate may be summarized as follows—

He must be able

1. To show theoretical knowledge and understanding of all the factual information relevant to his subject and skill in demonstrating this in a practical way.

2. To record everything in a concise form, with balanced and logical arrangement.

3. To illustrate it pictorially, adequately and relevantly, with clearly made and fully labelled diagrams or drawings.

4. To write legibly, grammatically and with correct spellings.

Nothing short of perfection in all these can be expected to gain the fullest awards, but "practice makes perfect" and if the reader ensures for himself practice by perusing old papers and selecting and answering questions in the manner to be outlined, and by setting himself practical exercises to be accomplished in a given time, he will have every chance of reaching the highest standard.

The examination is normally in two parts; written papers, usually two, each of three hours duration, and a practical examination of three hours. It is necessary to reach the required standard in both parts to obtain a pass.

In the following sections there will be outlined a procedure to be adopted in the written papers and this will be followed by hints on how to deal successfully with the practical paper.

ADVICE AND SUGGESTIONS FOR THE CANDIDATE

All those who have had the experience of marking examination scripts know that the great majority of candidates could have improved their performance had they followed some simple rules of technique. In the hope that some candidates may profit by adopting a system which has been tried and tested over many years, it is set out below.

The Written Paper

SELECTION OF QUESTIONS AND TIMING

All papers are carefully checked for errors and the candidate can assume without hesitation that, unless otherwise stated in an amendment, the questions are exactly as the examiner meant them to be, even though he, the candidate, might think that improvement could be made by altering the wording in various places or by construing parts of it according to his own mistaken fancies. All instructions, whether concerning the examination as a whole or as applied to individual questions are meant to be obeyed implicitly. *The ability to comply with straightforward instructions is part of the test of a good student.* Failure to do so, for any reason, indicates a weakness in the candidate who will in consequence be penalized by loss of marks.

On receipt of the question paper, the candidate should carry out the following preliminary procedure, taking about five minutes to do so. In some instances this time is allowed in addition to the stated examination time.

1. Check that the correct paper has been received. (In a mixed examination a mistake could occur.)

2. Read the heading, noting any particular instructions. Do not rely on the memory of what may have been stated in previous years. Examiners change periodically and a new examiner may differ in his requirements from those before him. Look especially for *how many questions are to be attempted* and *whether one or more is compulsory*. Note also that biology papers will also carry instructions in the heading concerning the use of diagrams. Remember always to bear these in mind.

3. Read *all* the questions *slowly and carefully*, making sure that any printed on the reverse of the sheet are not overlooked. In each case, indicate the possibility of an attempt according to whether there is confidence of the possession of a good knowledge of the subject matter relevant to the question. If at this stage a candidate has been able to tick confidently the full number of attempts required, any examination nerves he may have suffered will normally have receded. A deficiency of possibilities is almost always due to faulty preparation and it is too late to rectify such a condition. But whatever the situation, there must be no panic. A second reading of the paper and a little deep thinking may go towards the restoration of confidence, and there may be some consolation gained from the fact that attempts at parts of questions are not debarred and can earn a proportion of the marks.

4. Having selected the possibilities, and assuming that there is a surplus of these, fix firmly those which are to be the certainties, leaving the more doubtful prospects for later judgement, but be sure to undertake first those questions not only at which the best attempts may be made but also those which can be completed in the shortest time. When estimating time required, take into account— (i) can the question be answered by direct factual information calling for the least possible deep reasoning? (ii) Will there be need for many diagrams? If the answers to these are positive and negative respectively then the question may be considered "short." It is frequently the case that problems in Genetics are among the "shortest" of all, if the means of solution is arrived at quickly.

The reason for taking the shortest answer is a very sound one. Time is an all-important factor and any time which can be gained at the commencement of the examination is of great value later when it can be used to greatest advantage over those questions which may call for deep pondering and memory searching. The most important thing of all is not to be caught short of time, that is to say, with the full number of attempts not made. The reason is fairly obvious. If, say, four attempts are made instead of five, then the maximum mark which can be achieved, even if these answers are perfect, is 80 per cent. Perfection in an examination answer is a rarity, and an analysis of the average marks obtained by students in the biological examinations show that the majority of answers gain about 50 per cent or less. This means therefore that for many students, failure to attempt the full number of questions can very easily result in failure of the whole paper since 50 per cent of 80 is 40. It is more difficult and time-consuming to gain the second 50 per cent of the marks than the first 50 per cent on any one question and so it is much more likely that the average student will gain 50 per cent of 100 marks, i.e. 50 marks, by attempting five answers, providing of course that he is in a position to do this, than he will gain the same number of marks by achieving $62\frac{1}{2}$ per cent on each of four attempts. The safest way therefore is to make the full number of attempts, even if one or even two may have to be treated with greater brevity than might be deemed acceptable.

Assuming that five attempts are to be made in three hours, this gives at least thirty-five minutes per question and allows a few minutes in hand at the end. To assist in time estimation, an examination room clock is normally provided and this should be consulted at frequent intervals.

ATTEMPTING AN ANSWER

The following guide can be of aid only to those students who know sufficient of their subject to profit by its adoption.

1. Read the selected question once more, slowly and carefully, to ensure that the first impressions were not misconceived. Then *analyse the question* according to the following—

Every question or part of a question will have *two* significant words or phrases. One will be interrogatory or instructional, the other will indicate the biological matter about which the interrogation or instruction is made. There are generally words or phrases which qualify one or both of these. To make the analysis, *pick out and underline all the significant parts, grasp firmly in the mind their significance and mentally decide to obey them implicitly without the slightest deviation.* There is no need to write out the question, the underlining can be done on the question paper.

This is the only sure way of not misunderstanding the question and should take no more than a minute or two.

Example of question analysis

Make one labelled drawing to show the detailed structure of the ovule of an angiosperm just before fertilization. Describe concisely the changes which occur after fertilization in the development of such an ovule into a seed.

	Interrogation or Instruction	Qualification	Biological topic	Qualification
Pt. 1	Make drawing	One labelled	Structure of angiosperm ovule	Just before fertilization detailed
Pt. 2	Describe	Concisely	Changes ovule to seed	After fertilization

It is sometimes the case that a student cannot decide the precise meanings of certain instructional remarks and may thus do more or less than he should. Below is a list of the more frequently used terms with notes which should make their meanings clear.

| Give a description of | Describe . | Represent in words (with diagrams where these assist) to enable the reader to form an idea of an object, idea, incident, etc. |

Give a description of .   } Represent in words (with diagrams
Describe . . . . }   where these assist) to enable the reader to form an idea of an object, idea, incident, etc.

Give an account of   . . Write an explanatory description or narration.

Discuss .   .   .   . Expound the various views held upon, or the various factors to be considered.

Compare   .   .   . Put side by side, two or more things, to observe for similarity or other relationship to one another.

Contrast   .   .   . Bring out the differences between.
Compare and contrast . . Bring out the similarities and differences between.

Distinguish between   . . Bring out the essential features of things which makes each distinctive from the others.

Explain .   .   .   . } Make known in detail, make intelligible.
Account for .   .   . }
Indicate .   .   .   . } Point out, make known, make understood.
Show   .   .   .   . }
State   .   .   .   . Present in the form of a statement or considered words what is a fact, condition, problem, etc.

Define .   .   .   . State precisely what is comprised in or meant by.

List   .   .   .   . Enter in a catalogue, roll or inventory.
Summarize   .   .   . Make a brief account of, give a resumé.
Survey .   .   . } Make a cursory inspection of, make a
Outline .   .   . }   general account of (as opposed to detailed).

Enumerate	Count off, specify by items, usually in order of considered importance.
Review	Subject to examination or inspection.
Write an essay . . .	Write a personal appreciation of.
Comment on	Make explanatory remarks or criticisms upon.
Consider	Examine with a view to acceptance or rejection, weigh the merits of.
Criticize	Write a judgement of the merits of (especially pointing out faults).
Amplify	Enlarge upon, add detail to.
Exemplify	Give examples of.
Support your decision . .	Bring facts to confirm, substantiate, bear out or back up.
Classify	Arrange in classes or groups.
Emphasize	Lay stress on.
Correlate	Bring things into mutual relationship.
Illustrate by reference to .	Use specified example(s) to make meaning clear.

2. Having fixed clearly and precisely what is to be done, *make a rough plan of your attempt on the answer book.* This need only be very brief, by abbreviation, by symbols or any jotted notes. If it is later struck through, the examiner will not include it for marking purposes but cannot fail to be favourably impressed by a methodical candidate. In the plan include all points which must be considered or mentioned. *Carefully judge their relevancy and eliminate everything not vital.* No marks are awarded for irrelevant matter and to include it is time-wasting in the extreme. Arrange all points or facts in logical or balanced order. Fix the places at which diagrams can be included with best advantage. If the question calls for a diagram only, it is still worth while to make a preliminary rough try. When the plan is complete, reference to it from time to time will facilitate the composition and balance of the final attempt. To draw up such a plan should take no longer than five minutes.

> To go in drawing—2 integs., micropyle, nucellus, funicle, V.B., emb. sac, 8 nuclei—2 synergids, oosphere, 2 polar nuc., 3 antip. cells, remains of 3 potential megaspores, portion of placenta.
>
> Caption—L.S. anatropous ovule, just before fert.
>
> Description—Antipodals and synergids disappear—zygote develops to form basal cell, suspensor, embryo cell pushed deep into emb. sac. which enlarges at expense of nucellus.
>
> Endosperm nuc. (triploid) rapidly divides to form endosperm.
>
> Diagram up to this stage.

Embryo cell becomes embryo embedded in endosperm which completely replaces nucellus. If seed endospermic—no further development.

Diagram.

If seed non-end.—embryo enlarges, particularly cotyledons, endosp. disappears—food stored in cots.

Diagram.

Integument becomes testa.

Whole struct. dries out and becomes detached from funicle.

3. Systematically work through your genuine attempt according to your plan. If previously omitted relevant material comes to mind during this period, quickly jot it in on the plan if there is still time to include it in the correct sequence. If it is too late for this, it must be appended at the end of the answer, but if this is done, indication should be made in the answer at the point where the addition should have been placed. A time of about twenty minutes may seem to be short for this part of the work, but if the preliminaries have been properly carried out, there will be no necessity for great deliberation during it. It is surprising how many words can be written in long-hand in this time when the writer does not have to pause intermittently to decide what to write about next.

Above all, always make your meaning clear. Examiners are not mind readers and in any case cannot be expected to view favourably the candidate whose statements are so obscure as to call for great effort to make them intelligible. Write to the point always with short, sharp sentences. Make no attempt to pad or fill out the answer with flowery language or irrelevant matter. *No marks can be gained by so doing*.

Unless the question expressly calls for diagrams or drawings, it is left to the candidate's discretion to include them or otherwise. Since it is rarely the case that diagrams are superfluous in an answer to a biology question, and since the examiner almost always indicates at the head of the paper that, *large, clear, correctly labelled diagrams carry many marks*, a candidate will always stand to lose by failing to include suitably chosen pictorial illustrations. On the other hand, this must not be taken to mean that diagrams can always replace written description entirely, unless this is specifically stated in the question. Choice of diagrams and where to include them often constitutes one of the greatest worries but this can nearly always be lessened if during the planning of an answer these problems receive due consideration.

Every diagram included must be *adequately sized to show detail, clearly labelled, preferably in ink, must have a caption, and can be*

numbered if this makes it more easily referred to in the written work.
The most sensible equipment for drawing will include—

A soft pencil, three or four differently coloured ball pens, ruler and compasses.

Coloured wax pencils or crayons can be used but it is *obviously a waste of time to colour in large areas of paper when these can just as clearly be delimited by hard, sharp coloured lines.*

Diagrams of apparatus should always be neatly drawn, roughly to scale, with a ruler. Circles should be drawn with compasses.

4. When the attempt is completed, *read through it immediately.* Check carefully that, as a result of writing at high speed, no words have been omitted, particularly any which might alter the whole sense of a sentence or passage. Spelling and legibility must also come under review. This may take as long as five minutes but in many cases is time well spent.

When all the requisite attempts have been made there should still be about five minutes left for final checking that all instructions have been complied with, for instance, that each attempt has been correctly and clearly numbered, that the candidate's number is on the answer book, and so on.

The Practical Examination

This is not only a test of manual dexterity in the performance of operations relating to biological material and phenomena, but is also a great test of a candidate's general intelligence. Translation of book knowledge into sound practical demonstration calls for a high level of understanding, insight and reasoning power as well as technical skill. Since biology, like all other sciences, cannot progress without those skilled in the manipulation of the relevant materials, it is necessary that a student should at this level indicate that he has reached a certain standard of ability to correlate "theory" with observable or demonstrable fact, that he can plan simple experimental investigations and that he possesses the necessary practical skill to accomplish these, before proceeding to study at higher levels.

All practical examinations, whether in Botany, Zoology or Biology, at this level, tend to conform to the same pattern. The contents of papers set over many years by different authorities may be analysed broadly as follows—

1. Tests of ability to recognize, display and record observations on morphological, histological and anatomical features of plants and animals. These include—

(i) Section cutting and simple staining of plant parts, microscope examination of these preparations and interpretation of what is is seen by means of clearly labelled drawings.

(ii) Dissection of a whole or part of an animal organ system and the interpretation of the anatomy displayed by means of a clearly labelled drawing.

(iii) Recognition and graphic illustration of the external morphology of whole or parts of plants and animals.

(iv) Classification of plants and animals with reasons.

2. Exercises in the manipulation of organisms, reagents, apparatus, etc., usually involving an investigation of some simple physiological phenomenon and test of ability to make and record accurately all methods and observations. Such exercises might embrace osmosis, permeability, enzyme action, food storage, etc.

3. Exercises in the interpretation of observations. These can be set in the form of instructions to comment on an experiment or demonstration which has been set in operation by the examiners.

4. Tests of technical skill in making microscope preparations of a simple kind.

5. Exercises in identification of whole or parts of plants or animals, with or without reasons for the identification.

6. Exercises involving the making of comments on the biological significance of features to be observed in special material, for example, morphological or anatomical adaptation to habitat conditions, special mechanisms of pollination, dispersal, etc.

7. Identification of an angiosperm using a flora.

8. Exercises in comparative morphology and anatomy, particularly floral morphology, animal parts, especially skeletal, brains, etc.

It is not intended here to give practical details of how to answer specific questions but only to guide the student on how to treat the examination as a whole, particularly from the point of view of time allocation, and in general terms, how to deal most efficiently with each type of exercise outlined above.

Preliminary Preparations

In addition to the necessary mental equipment it is essential that the student be adequately materially supplied. The examiners will always supply items of apparatus, biological material, reagents, stains, etc., but the student is instructed to take with him a set of biological instruments, pencils, eraser, pen and ink, a clean duster and sometimes a flora. Instruments should be carefully inspected shortly before the examination day and repairs and replacements made as necessary. For example, it is quite impossible to cut good sections with anything other

than a perfectly sharpened razor; it is equally impossible to perform
fine dissection of any kind without good scalpels, fine forceps which
meet at the points and fine scissors which cut to the points. All these
things must be checked and attended to in good time. Pencils must be
sharpened ready for use and pens filled with ink. Coloured pencils or
ballpens can sometimes be used with advantage but should not be used
indiscriminately, particularly when making drawings of animal dissections and sections of plant parts.

The flora when required, must be a publication acceptable to the
examiners.

Should a student require to make use of any reagents or stains not
supplied at his bench space he is at liberty to ask for them. This does
not necessarily mean that they will be supplied. The examiner may
consider his allocation to be perfectly adequate, which of course, it
usually is.

One preliminary piece of preparatory work can be carried out only
in the examination laboratory; that is the checking and adjustment of
the microscope provided. Microscopes provided for use by examination
candidates are notorious for their state of dilapidation. Nevertheless,
they would not be there if they were not capable of giving the required
service. This they cannot do if they are not correctly set up before use.
Check the light source, the mirror, the opening and closing of the iris
diaphragm, that the condenser is adjusted for critical illumination, and
that the optical parts are clean.

The Practical Paper

SELECTION OF QUESTIONS AND TIMING

Most of the opening remarks concerning the written paper apply
equally well to this, but certainly not those relating to choice of
questions. Almost invariably there is no choice. Candidates are expected to attempt all exercises. Generally, only the very best of
students can, in fact, achieve a very high attainment in all the exercises
in the time available, usually three hours. Speed and accuracy are
absolutely essential. It is therefore necessary that the average student
should use his time to the maximum advantage. In order to do this he
is well advised to study the following comments carefully so that when
he sees the question paper he can quickly assess the relative mark
values of the questions and plan his order of working and the time to be
allotted to each exercise accordingly.

Practical questions do not all carry the same marks and in general the
marks are allotted in proportion to the time necessary to complete
the task reasonably well. For example, an animal dissection and
drawing, or a plant anatomy section, its interpretation and drawings,

cannot be completed in anything under an hour each. Hence such a piece of work must carry at least one-third of the total marks, more often as many as 40 per cent. On the other hand, straightforward identification of "spots" may take no more than a minute or two each at the most. Hence they will be assessed on a much lower scale. The performance of a simple physiological investigation may take upwards of half an hour to complete but does not necessarily need full attention all the time; it can be initiated and returned to at intervals throughout the examination time. Thus this may not carry more than about one-quarter of the total marks. A simple temporary microscope preparation can be completed in less than fifteen minutes even at a slow pace and thus cannot be regarded as meriting a high score however well it has been made. A stained, permanent preparation of animal tissues may take up to half an hour and may carry 20 per cent marks.

Generally speaking, it will be the animal dissections, plant histology and anatomy exercises, physiology and floral morphology work which carry the highest proportions of marks. Even in these, perfection is very elusive and to spend more than seventy-five minutes on any one of them is unlikely to be profitable from a mark-gaining point of view, particularly if another exercise has to be done very hurriedly or omitted completely. The student is advised to get to work swiftly on one of these high mark-carrying exercises for a time not greatly exceeding one hour and then turn to other questions, being prepared at any time to divert attention to any physiological exercises he may already have set under way. Should time be available later the work can easily be returned to and additions and improvements made to dissections, sections and drawings.

The problem of what to attempt first always arises. This really can only be solved after the question paper has been carefully digested and any advice given here must be weighed accordingly. Plans of attack for given sets of circumstances can be worked out previously if old examination papers are studied and each is treated as an exercise in work timing and arrangement.

If an exercise involving a physiological investigation is set, requiring examination of material at intervals over a long period, then that should be initiated immediately. Then the major task should be worked upon for a total period of not less than an hour nor more than an hour and a quarter, interrupted if necessary for the examination of material undergoing treatment already. If no such long-drawn-out exercise is set, then it is best to concentrate from the start on the major questions.

It is sometimes the case that the student may be called away from his work to examine some demonstration for a limited period, or he may be allowed to retain the "spot" specimens for a few minutes only, at a time specified by the examiner. If this is so, then the student has no choice and must make full use of his allotted time on those exercises. If this is not so, then such short tasks are best tackled when the bulk of the more protracted work has been completed, but should never be omitted simply because the student wishes to polish up a dissection or drawing.

Time involved in drawing, recording observations, writing notes and indicating identities can be shortened if at the outset it is decided exactly where in the answer book the questions are going to be dealt with and the necessary space allowed. Any method sequence or periodic observations can then be entered in chronological sequence without the usual rough paper jottings which not only are not good scientific practice but cause duplication of effort. Similarly, drawings once started can be added to and improved at any time. Identifications and other necessary notes can be filled in on the appropriate page as they are arrived at, and so on.

Remember always that greatest efficiency and economy of effort can be achieved only as a result of intelligent planning, no matter how simple the task may appear to be.

How to Deal with the different forms of Practical Exercises

1. PLANT HISTOLOGY, ANATOMY AND DRAWINGS

Typical form of question—

> Cut a transverse section of the material provided, stain it appropriately and mount it in glycerine. Draw a Low Power plan of the tissue layout, fully labelled, and a labelled High Power cell drawing of the vascular tissues. Leave your section on the microscope stage for inspection.

Note—

If the examiners provide a permanent preparation, the early part of what follows will not apply.

Section Cutting

It is essential to obtain a thin section cut in the transverse plane. Thin sections can be cut only with a really sharp razor. A blunt razor will tear or cut thick slices only. Truly transverse sections, i.e. not oblique, can be cut only if the material and the razor are held correctly. Much practice is necessary before it becomes a matter of routine to produce good sections and it is hoped that the student will by this time be proficient.

In the examination, embed the material in pith (if necessary) and proceed to cut sections continuously for several minutes, and then with a black-and-white tile as background, select for further treatment the three or four which are obviously the thinnest. Remember that a full cross-section is not always essential since the part may be radially symmetrical, in which case a very thin wedge of only a part of the plant organ will do. At the thin end of the wedge, cell detail usually shows up perfectly.

Staining

Unless a particular stain is asked for, the student can select his own from those provided. He should be familiar with the uses of Acidified Phloroglucin, Aniline chloride, Aniline sulphate, Schultze's solution (Chlor-zinc-iodide), Methylene blue, Iodine, etc., as stains for temporary purposes.

Mounting

Unless specially asked for (very unusual), do not attempt to make a permanent preparation involving dehydration and mounting in Canada Balsam. This is a waste of time and cannot gain any extra marks. It indicates clearly to the examiner that the student is quite unable to follow simple instructions and he may even be penalized. Use glycerine after washing the stained sections quickly in water. Mount all the stained sections. During this operation remember the rules—clean slide and cover glass; sections centrally placed on slide and not overlying one another; not too little, not too much mountant; no air bubbles; no mountant on coverslip; blot off excess mountant on slide if necessary; If mountant gets on the coverslip, discard and remount, it is impossible to see through this. If the first attempt is really poor, mount all over again; it is usually fatal to try to improve a poor mount by poking and prodding it.

TIME SO FAR—10–12 MIN.

Microscope Examination

Check that the microscope is properly adjusted. Place Low Power objective in position and slide on stage, clipped down. Examine all sections on Low Power and select the best for further attention. Study it in detail on both Low and High Powers to check that it does show what is required. Try adjusting the diaphragm aperture to get the best illumination for detail.

TIME SO FAR—15 MIN.

Drawings

(i) L.P. plan or map of tissue layout.

This means exactly what it says. Draw to the largest possible scale, the whole or a representative part of your section, plotting within the outline the positions of all the different types of tissue which you recognize. Check carefully the relative proportions of each tissue, numbers and arrangement of vascular bundles, etc. *Do not draw a diagram: do not attempt to draw cells nor waste time colouring large areas of paper.*

Label correctly the different *tissues* which you have recognized, e.g epidermis, collenchyma, sclerenchyma, parenchyma, protoxylem, meta-xylem, protophloem, metaphloem, endodermis, pericycle, starch sheath, vascular cambium, cork cambium, cork, phelloderm, etc. *See* Fig. 3.

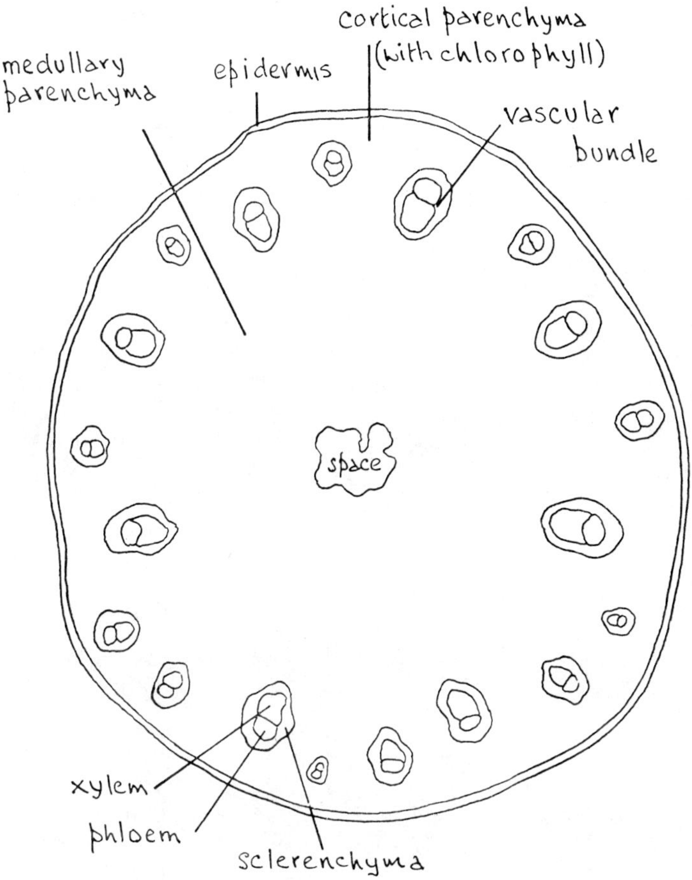

Fig. 3

Time so far—30–35 min.

(ii) H.P. cell drawing.

In this must be included the detailed structure of the types of cells asked for by the examiner. They must be drawn in their proper positional relationships and proportionate sizes to one another, not as isolated patches on the drawing paper. No cell should ever be drawn without either some adjacent cells or parts of same. Where it is necessary to draw cells from adjacent tissues, draw also the region where the two forms are contiguous or blend into one another. Wall detail must be shown indicating proportionate thickness of the wall, laminated

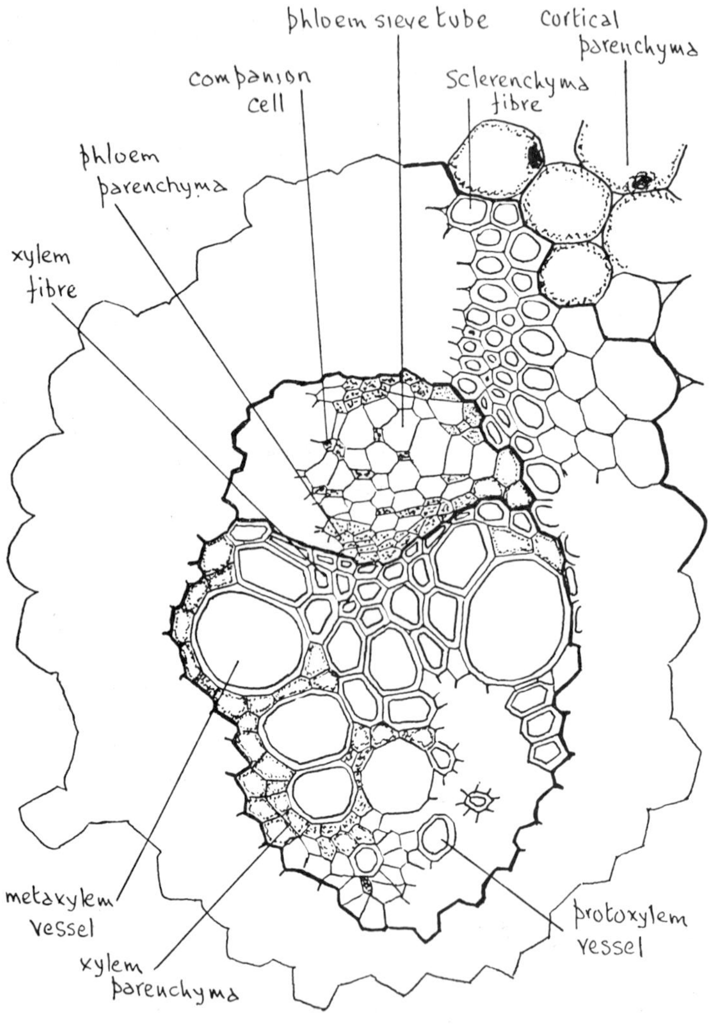

Fig. 4

appearance and pitting. If walls are comparatively thick then a double line must be used to represent the thickness. If walls are thin then single lines can be used to represent them. Any attempt to represent thin walls with a double line usually ends up with the wall thickness completely out of proportion to the cell size. If cells have living parts, i.e. cytoplasm, nuclei, or contain non-living inclusions, e.g. starch grains, crystals, etc., these must be drawn. All lines must be clear-cut continuous pencil lines; impressionist sketches will not do. Do not colour.

Commence drawing by outlining lightly on as large a scale as possible the whole area from which your cells are to be taken. Within this outline draw in the areas which contain your cells. In other words draw quickly a much enlarged L.P. plan of a small portion of your section. (This is indicated by heavy lines in Fig. 4.) Now draw the relevant cells by first outlining the *outside* faces of the cell walls, then represent wall thickness by a line inside this (if the wall is thick enough to warrant the double line). Any attempt to draw first the outline of the cell cavity is usually disastrous and constitutes what is known as "drawing holes in the paper." Good accurate cell drawings can be made only by the student who really understands their nature.

It is generally accepted that the smallest cell drawn should not be less than one-quarter of an inch across its narrowest part. This can sometimes lead to difficulties if very large vessels are to be drawn in proportion to, say, very small cambium or phloem cells. A big vessel may need to be as much as two or three inches across, which makes the paper usually provided rather too small. In such a case, expand on to another sheet and explain that the two pieces must be placed together if the drawing is to be studied in one piece.

Continue to draw cells for about half an hour. It has been suggested that an average student should be able to draw in detail about fifty cells in this time.

When the drawing is complete, label every type of cell shown, e.g. metaxylem vessel, protoxylem vessel, tracheid, xylem fibre, xylem parenchyma, sieve tube, companion cell, phloem parenchyma, cortical and medullary parenchyma, collenchyma cell, sclerenchyma fibre, stone cell, etc. *See* Fig. 4.

Remember that the work can be returned to later if time allows and additions and improvements made.

Total time—about 1 hour.

Sometimes a specimen of stained, macerated tissues is provided from which the candidate is expected to isolate and make drawings of cells of particular tissues. In this case the sample should be transferred from its container into a watch glass of dilute glycerine and lightly teased with brush and needles until small portions are floated out. These should then be mounted and searches made until good specimens of the required cells are found. These should be drawn only after close examination with the High Power, using varying light intensities to give best results of detail. Pits in walls, primary, secondary and tertiary wall layers, cell inclusions, etc., must be searched for and shown in the drawings where applicable.

2. FLORAL MORPHOLOGY AND DRAWINGS

There is no standard form for this type of exercise but the candidate is usually expected to carry out one or more of the following—

By means of a flora, identify the specimen. From a dissection of the flower construct a fully labelled L.S. drawing or a labelled drawing of a half flower; construct a floral diagram; compose the floral formula; give a written description of the flower; comment on possible pollination mechanisms; refer specimen to family, with or without reasons.

Identification with Flora

There is little useful advice which can be offered. The student must be thoroughly familiar with his flora and have had adequate practice in its use. Correct identification results from accurate observation of both vegetative and floral structure. All relevant observations should be recorded. In the event of wrong identification the examiner is then able to see where the error was made and judge accordingly.

Careful dissection of the flower is always necessary and should be carried out as below.

Dissection of Flower

After a careful examination of the whole flower, the parts should be removed and arranged in their correct relative positions as far as this is possible. Parts must be severed or split open with a sharp scalpel, handled with forceps and always examined under the hand lens. In a case where all the parts are freely inserted in whorls, the task is comparatively easy. When parts are joined, as for example to form calyx and corolla tubes, they should be removed as complete units, slit open along either the anterior or the posterior margin and opened out. If the stamens are epipetalous they will be displayed at the same time. Dissection of a very small ovary is sometimes difficult. The whole specimen should first be searched for a ripe ovary (fruit) from which the structure can most easily be observed. In the absence of a mature ovary, it may be essential to section a very young specimen with a

standard petal
wing petal
stigma
style
stamens
keel petal
ovary
sepal
receptacle
a.
b.

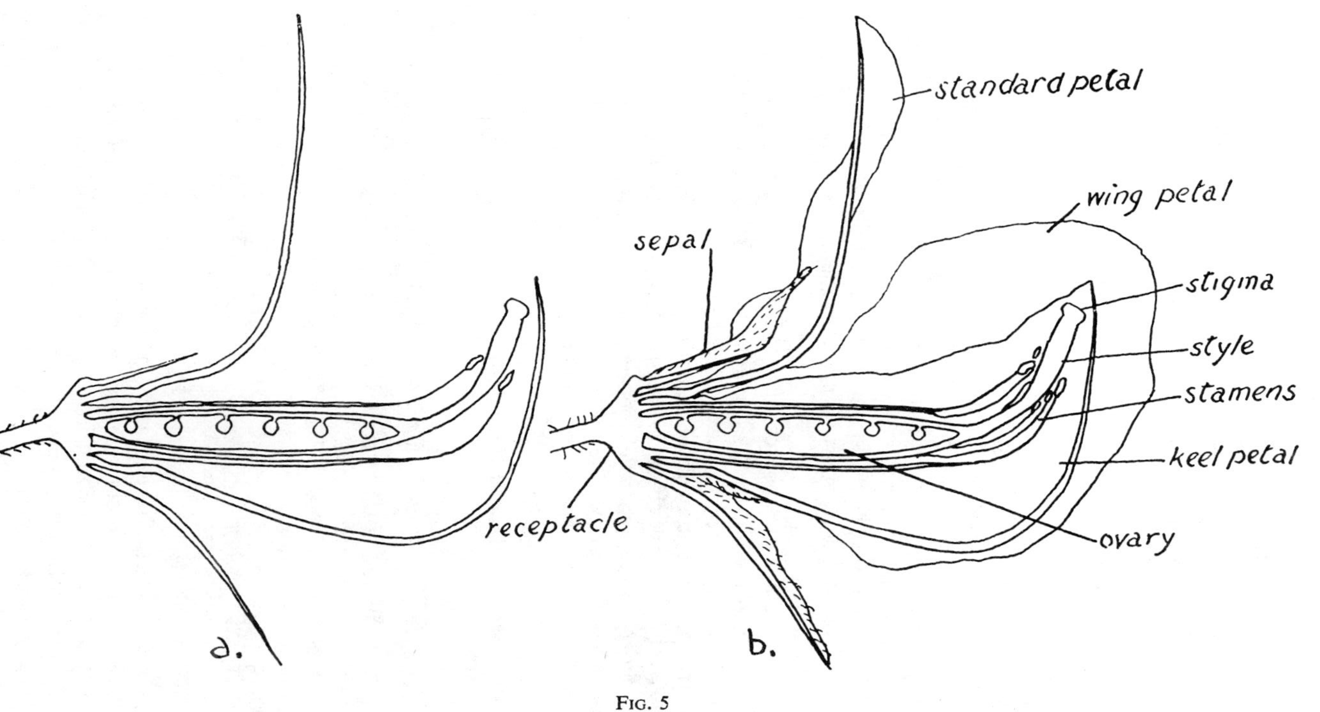

Fig. 5

razor and examine it under the low power of the microscope. As the dissection proceeds, notes of all observations must be systematically recorded.

From the notes made during the dissection, any of the remaining exercises can be speedily completed.

TIME TO DISSECT, RECORD AND IDENTIFY—10–15 MIN.

Drawings

A L.S. of a flower must show only the surfaces of the parts which would be cut by the razor if the whole flower were embedded and the thinnest of sections were taken on the microtome in the plane of the anterior/posterior axis. The half-flower must show these cut surfaces and the other whole floral parts which would be seen in the background if the flower were simply cut in half in the same plane (*see* Figs. 5 (*a*) and (*b*)). These drawings must be made to the largest scale which the paper will allow.

A floral diagram must be constructed on a large scale, based on circles drawn with compasses. The innermost circle, within which the ovary is drawn, must never be less than one inch in diameter, otherwise carpel arrangement and ovule placentation cannot be shown accurately and in detail. A cross check between a L.S. and the floral diagram should show no discrepancies between the drawings. If the floral diagram is bisected across the anterior/posterior axis, the line will cut the parts the surfaces of which should be showing in the L.S. No other parts should be showing in the L.S. drawing (*see* Fig. 6).

TIME TO DRAW L.S., HALF FLOWER AND FLORAL DIAGRAM—ABOUT 5–10 MIN. EACH.

Floral Formula

Knowledge of the floral structure and the meanings of the symbols customarily used is essential if the formula is to be composed accurately. It can be composed very quickly from the notes made during the dissection of the flower.

TIME—2–3 MIN.

Pollination Mechanism

The candidate will be expected to comment intelligently on the possible mechanism, basing his comments on an understanding of floral adaptations to wind and insect (or other) pollination. He must draw attention to the presence of these in his specimen and make any necessary drawings.

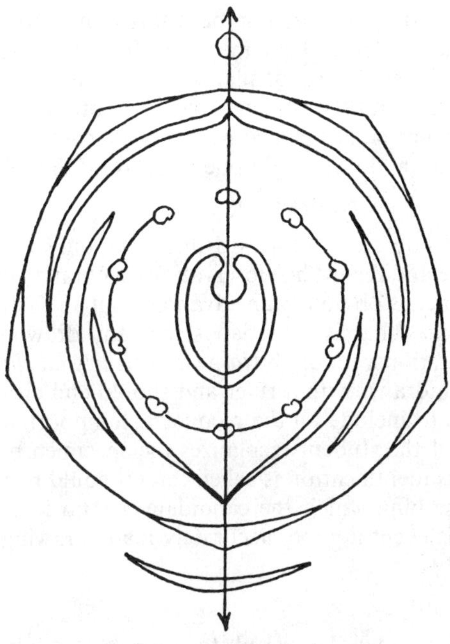

Fig. 6

Time—5–10 min.

3. Plant Physiology Exercises

These are usually simple experiments designed to test the candidate's ability in several directions. These include his ability to obey instructions intelligently, carefully and implicitly, his ability to make accurate observations, his ability to record all methods and observations concisely but adequately, and to draw valid conclusions.

Examiners expect that experimental methods will be fully recorded even when a good deal of instruction is given on the question paper. Observations must be tabulated or graphically expressed where necessary. Any conclusions must be clearly set out and the reasoning on which they are based given in full. If the candidate complies fully with these requirements, he can be sure of gaining a good proportion of the marks even though the observations may not be wholly accurate and the conclusions incorrect.

4. Plant "Spot" Slides and Specimens and Identification

These may be selected from a very wide range. In the general case they will be material of a kind which a student could be expected to have encountered, although not necessarily any of the actual types specified in the syllabus. They may vary from whole plants to microscope preparations of parts of plants. It is impossible to forecast the nature of "spots" and the student must rely entirely on his intelligence and the experience he has gained during practical work. The examiner may require the candidate to do one or more of the following—

(i) Make labelled drawing(s).

This is perfectly straightforward and asks only that the candidate should observe and record his observations accurately in pictorial form. Skill with a pencil is obviously an advantage but an "artist's impression" is of no value. What is essential is that any drawing(s), as with all pictorial illustrations, should be *large* and *clearly labelled*. All labelling lines should be drawn with a ruler and should end clearly on the parts labelled, these to include all the recognizable important features. It is an advantage if the student recognizes the specimen but this is not an essential unless identification is asked for. It could be the case that the "spot" is something which the candidate has no knowledge of whatsoever. In this event a good, accurately made drawing alone will gain some marks.

(ii) Identify.

The candidate may be asked to support his identification with reasons. If this is not the case then the identification only should be written. Fullest possible identification should give which part of which plant or which whole plant down to species, but this is frequently a difficult matter even for a specialist. Examiners rarely expect candidates to identify to specific level or even generic level. A candidate would be expected to place a flower of the commoner families in its family, but in general the identification required is of the form—"A second year female cone of a gymnosperm of the Pinus group. Division Spermatophyta," or "The sporophore of a member of the Agaricales. Division Eumycophyta," or "A gametophyte of the class Hepaticae. Division Bryophyta," or "A coenocytic, filamentous alga of the Chlorophyceae. Division Chlorophyta." If the specimen were a section of a plant organ, then the candidate would be expected to recognize the organ and the plane in which it had been cut, i.e. T.S., L.S., etc.

When it is necessary to support the identification with reasons, the only valid ones are those based on the recognition of the features characteristic of the group named, i.e. it is a thallophyte because . . .; it is an alga because . . .; it is a green alga because . . .; it is a coenocyte because . . .. The student is well advised to be familiar with the characteristics of the main plant groups. Should there be any difficulty in even roughly placing the specimen it is often a help to proceed by a process of elimination, working through the classification of plants in broad outline until a sensible attempt can be made. A wild guess which almost certainly will be wide of the mark can gain no marks and may cause the examiner to doubt seriously the intelligence of the candidate.

      (iii) Comment on features of biological significance or importance.

Almost invariably such specimens clearly exhibit adaptations or modifications to be associated with particular conditions. It is a matter for the candidate to relate the observed adaptations to the supposed conditions as clearly and concisely as possible. Even if drawings are not specifically asked for they can often help in clarifying a meaning or intention, but labelled drawings alone would not necessarily constitute a fully satisfactory answer.

It is most important that the candidate should remember always that he cannot do his best with any "spot" material until this has been examined in the fullest possible way, by dissection or sectioning and hand lens or microscope examination, wherever these apply.

5. ANIMAL DISSECTIONS AND DRAWINGS

Animal dissections are of two general types, either a particular system (or part of one) or a general dissection to display as much of the anatomy as possible.

Preliminary Work (Preparatory)

Decide quickly whether the dissection should be begun dorsally, ventrally or laterally and fasten the animal in the dissecting dish or dissecting board in the most suitable position. Ensure that the position, especially if in a small dish, will allow free play of the hands and instruments. The pins or spikes used, should be the smallest compatible with adequate support. For small specimens, e.g. cockroach, earthworm, the pins should be placed with their heads leaning away from the animal at an angle of about 45°. If the dissection is to be made on a preserved specimen, remember that the tissues will not stretch as much as in a freshly-killed animal. Dissect small animals under water; this will give the organs support and keep them more supple. When the animal is suitably fastened and you are ready to begin dissection, lay out your instruments in correct order, scalpels and scissors on the right, forceps on the left, needles and seeker in front.

The Actual Dissection

Get the preliminary skinning or opening the body cavity done quickly, and avoid cutting away large pieces or organs which will not interfere with the dissection. Portions which have to be cut away should be placed in a separate container or piece of paper; do not clutter up the dish and the water. Nothing is more annoying than having to sweep away small portions of tissue which drift across the dissection. In any particular dissection, the structures to be displayed should have some relationship to contiguous structures and hence wholesale removal of organs and tissues is not required. Cut away that tissue only which obscures those parts which should be displayed.

When dissecting very small structures, do not cut away hopefully but use a dissecting lens or a watchmaker's eye-glass.

Dissect along blood-vessels, nerves or ducts, and clean them of connective tissue, even though they may be very obviously visible. Unless otherwise instructed, trace nerves and blood-vessels to their points of origin or termination, e.g. cranial nerves to the brain, major arteries or veins to the heart. Dissection of an alimentary canal means along its whole length, and should include display of accessory glands with their ducts, the points of entry of the ducts, and the internal surfaces of those parts which have special significance.

Aim to display the main essentials of the dissection first, then do the major part of the drawing. Further detail of dissection can be carried out as time allows.

Display of the Dissection

Aim to display the specimen to show as much as possible of the system required. Judicious arrangement of pins is important, since you can show in your drawing, only those structures actually visible without moving anything. Entrances of ducts, etc., can be usefully indicated by pushing a bristle into the duct and cutting it short enough to be clear but unobtrusive.

Remove all scraps of tissue from the dissection, and cover with clean water, if in a dish.

Time Allowance for Dissection

Usually the major part of the dissection should be completed in one hour. A drawing is normally required and this will take about half-an-hour. It may be possible to add further details to both dissection and drawing at a later stage.

Drawing of the Dissection

Ensure that you have left enough time to do justice to this part of the work; normally at least 30 minutes will be necessary. You are usually asked for a *drawing*; this means that you must give as faithful a copy of the dissection as you can. Such a drawing should be a picture in a frame, the frame being the edge of the paper, and the clear border being utilized for labelling. Hence make your drawing large, centre it on the page and have adequate room for labelling. State roughly the magnification of your drawing, e.g. urinogenital system of male frog ×4, cranial nerves of dogfish ×2.

It may be necessary to draw an outline of the whole animal, e.g. general dissection of a cockroach or crayfish, but if this is so, details of appendages are quite unnecessary. Often, the dissection is confined to one relatively small portion of an animal, e.g. cranial nerves on one side; in this case only the part of the animal which has been dissected need be drawn.

Aim at correct shapes, correct positions and correct proportions. Often graduated callipers or at least a ruler will help here.

The drawing is a test of ability quite apart from the dissection; it will test your skill in transferring practical results to paper, and it will test your knowledge of the names of the parts. Therefore, even if you have made a bad dissection, draw it faithfully, and label it as fully as you can. If you have broken a blood-vessel, nerve or duct, indicate the break on your drawing. Before you start labelling, clean up the drawing so that there are no ragged or sketchy lines. If there are many structures of different nature which might be likely to cause confusion, some form of simple shading may be used, but there will not be time to give artistic impressions of solidity. Finally, all major *line* structures, e.g. ducts, nerves, and blood-vessels should be drawn in double lines, and if both arteries and veins are to be shown on the same dissection, it will help if you colour the arteries lightly with red pencil and the veins with blue.

Labelling the Drawing

In general it is perhaps preferable to use fine ink lines for labelling lines and also ink for the labels. All labelling lines should, as far as possible, be horizontal though at the upper and lower ends some may be vertical. Avoid oblique lines, and above all avoid crossing two or more labelling lines. If there is considerable detail, use top, bottom, right and left borders of the paper. Do not underline your captions and ensure that they are quite legible.

An example of a drawing correctly labelled is given in Fig. 7.

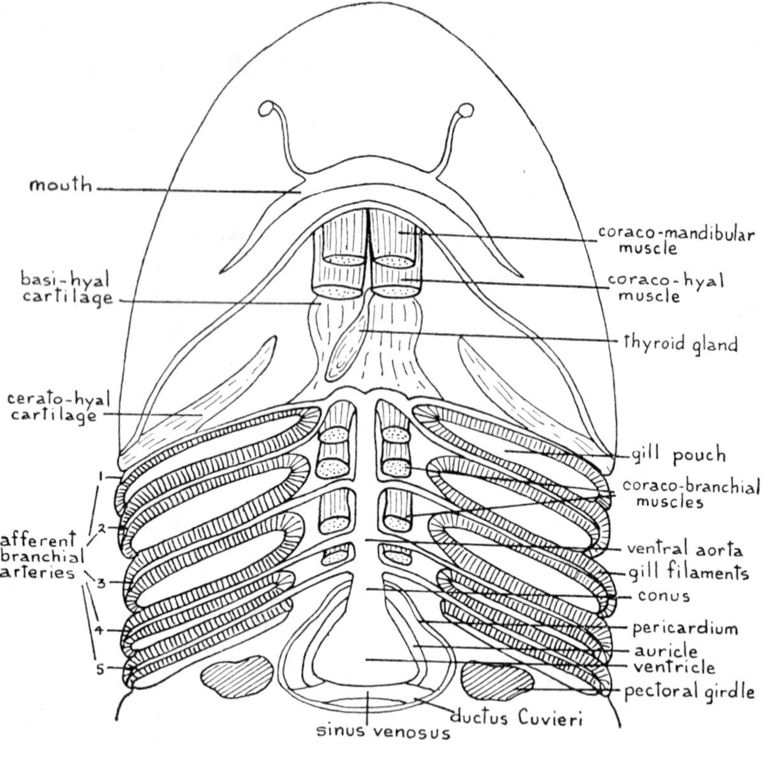

Fig. 7

6. Animal Microscope Preparations Permanent (or Temporary)

Your preparation will be marked on the following points—

1. Whether you have the correct tissue, animal or part of an animal.

2. Whether you have enough of the material to show its nature completely.

3. Well stained and differentiated (sometimes, staining is not required, e.g. mouth parts of insects).

4. Whether the preparation is properly fixed, dehydrated and cleared.

5. Whether the specimen is correctly mounted, in the centre of a clean slide and positioned correctly.

6. Whether the correct amount of mountant has been used.

7. Neatly and carefully labelled—usually only your examination number is required.

Method

1. Set out the fixative (if required), the stains, dehydrating, clearing and mounting agents in a row in front of you.

2. Have ready, clean slides, cover glasses, black-and-white tile, labels, and any instruments required. See that a clock is available or that you have a watch for timing.

3. Decide whether you will do the preparation on the slide, in covered staining dishes or watch-glasses.

TIME—4 MIN.

4. Ensure that you have the correct tissue or part, and enough of it to show the complete structure. Then make your preparation, i.e. smear, teased tissue, sectioned tissue or organ, stretched sheet, whole organ or part of animal, etc.

TIME—2–4 MIN.

5. Fix, if necessary (preserved animals are already fixed), and then remove fixative.

6. Stain for the correct time, agitating periodically. Remove excess stain; blue if necessary.

TIME—5 MIN.

7. Dehydrate (agitating periodically) and differentiate if necessary. Where the preparation is histological, ensure that nuclei stand out clearly.

8. Counterstain with a cytoplasmic stain if required. Then remove counterstain and blue again if necessary.

9. Complete dehydration and clear, ensuring that all water is removed. If not, revert to further dehydration.

TIME—8–10 MIN.

10. Test the viscosity of mountant and decide how many drops required. Arrange the material in correct position in centre of the slide; put on the mountant; lower cover glass slowly and then label neatly.

TIME—3–5 MIN.

TOTAL TIME—ABOUT 30 MIN.

7. DRAWING OF ANIMAL SPECIMENS

Drawing of a whole animal or part of an animal, a section or histological preparation.

1. In this type of question the student may be required to give labelled drawings of whole animals, e.g. *Hydra*, part of an *Obelia* colony, crayfish or insect, etc., a limb, a skull or other skeletal part,

a section through the body of an animal, or an embryo, or any kind of histological preparation. It is another method of testing knowledge of morphology, anatomy, histology or embryology and of the ability to make faithful recording of essential detail.

2. The larger the drawing, the more accurately can details be represented; use the maximum size possible on the paper and give some idea of the magnification.

3. Orientate your drawing—dorsal, ventral, right, left, anterior, posterior, proximal, distal—whichever apply. All lines must be thin, clear and firm, and proportions should be correct.

4. If the specimen is histological, draw a representative portion on low power, then a few cells on the highest power available. Do not shade unless there is likelihood of confusion, when a simple method such as stippling or hatching may be used.

5. If the question states *drawings* then it must be interpreted properly, i.e. for a whole specimen, dorsal, ventral and side views may be necessary, and for a section, anatomical and histological drawings may be required.

6. Label carefully and completely, with horizontal or vertical lines in ink, and clear legible names.

7. Do not identify, unless required to do so; you may be wrong.

8. IDENTIFICATION OF ANIMAL SPECIMENS (WITHOUT REASONS)

Large specimens, e.g. limbs, girdles, separate bones, whole animals large parts of animals, thick transverse cuts, etc.

Identify as accurately as possible but do not guess that it is the actual species you have studied. For skeletal parts state whether it is a whole skeleton, a limb, or a particular region, and whether it is made of bone, cartilage or chitin. For paired structures, state whether it is the left or right. Is it adult size or not, and can you state the sex? You are not asked to classify, so give the identification clearly and concisely.

Example

The left bony femur of a small mammal; a thick transverse cut through a cartilaginous fish, the anterior cut passing through the heart, and the posterior cut through the stomach; a typical cervical vertebra of a small mammal, etc.

Small specimens on slides or in tubes, e.g. protozoa, small coelen- terates, sections of embryos, small arthropods, organs or parts of organs, small appendages of arthropods, etc.

State whether it is a whole animal or part of an animal, a smear, teased or stretch preparation, or a section. If it is a section, consider carefully what type it is; the following list may help you.

T.S. = transverse section; this is at right angles to the long axis e.g. T.S. earthworm, *Hydra*, intestine.

V.S. = vertical section; this is at right angles to the surface at the point of cutting, e.g. V.S. skin, stomach wall.

S.L.S. = sagittal longitudinal section; this cuts the specimen into equal right and left halves—it is rarely perfectly accurate e.g. sagittal L.S. gastrula of a frog.

H.L.S. = horizontal longitudinal section; this is parallel to the dorsal (or ventral) surface, usually cutting the specimen approximately equally between dorsal and ventral, e.g. H.L.S. head of tadpole.

R.L.S. = radial longitudinal section; this is parallel to the long axis and passes through the centre; it is only used of radially symmetrical structures (common in Botany), e.g. R.L.S. of *Hydra*.

T.L.S. = tangential longitudinal section; parallel to the long axis but not passing through the centre (common in Botany); e.g. T.L.S. of a sea-anemone.

State whether it is stained or unstained; through what regions of the body it passes, and the name of the group which contains the animal if you cannot identify the actual animal (unlikely).

Examples

1. Stained specimens of ciliate protozoans, some showing stages in binary fission.

2. Stained transverse section of a small oligochaete annelid passing through the intestine.

3. A stained smear preparation of mammalian blood.

4. A stained sagittal longitudinal section of a completed amphibian gastrula.

5. A stained teased preparation of medullated nerve fibres.

9. IDENTIFICATION OF ANIMAL SPECIMENS (WITH REASONS—CLASSIFICATION)

First identify as shown in the last section. The reasons depend on your knowledge of classification, anatomy, histology, embryology, etc. Even though you know the specimen quite well, you must identify it from features which are present and visible.

(*a*) Place it in its correct phylum (and sub-phylum).

(*b*) Class within the phylum.

(*c*) If you can do so with certainty proceed to order, family and genus. Occasionally you may be able to identify the species.

Examples

1. Stained specimens of ciliate protozoans, *Paramecium spp.*
Reasons—

   (*a*) Adult animals of microscopic size, non-cellular—Phylum
   —*Protozoa*.

   (*b*) Possession of cilia, mega- and micronuclei—Class—*Cilio-phora*.

   (*c*) Cilia present in the adult phase—Sub-class—*Ciliata*.

   (*d*) Uniform ciliation—Order—*Holotricha*.

   (*e*) Free swimming, asymmetrical, gullet at the side—Family—
   Paramecidae.

   (*f*) Slipper-like shape; Genus—*Paramecium*.

2. Stained transverse section through a neurula stage of an amphibian embryo, trunk region.

Reasons—

   (*a*) Presence of notochord and dorsal hollow nerve cord—
   Phylum—*Chordata*.

   (*b*) Small diameter of notochord, neural groove—Subphylum
   —*Vertebrata*.

   (*c*) Absence of embryonic membranes—Class—*Amphibia*.

Without very specialized knowledge it is not possible to go any further with this specimen.

MODEL ANSWERS

Question

Compare and contrast, with respect to nutrition and reproduction, a *named* parasitic fungus and a *named* saprophytic fungus.

(*Note*—

This form of question is often regarded as among the most difficult. It is perfectly straightforward for the candidate who knows the facts and understands their biological significance. There is only one pattern of answer which is fully satisfactory; it results from an intelligent analysis of the relevant facts. All the facts must first be marshalled under headings relevant to the question and the items under comparison or contrast. For each such heading a statement must then be made including the facts and drawing attention to the similarity or difference between the two items with respect to these. The logical marshalling of the facts can be done as the rough plan; the second part constitutes the final attempt and can be written directly from the plan.

To write or tabulate two totally unconnected lists of facts, often disjointedly, indicates lack of ability to think biologically and can gain relatively few marks.)

Plan

	Parasite	Saprophyte
Nutrition Mode	*Phytophthora infestans* Heterotrophic (give meaning)	*Mucor mucedo* Heterotrophic (give meaning)

Kinds of nutrients required	Carbon source—not starch
	Nitrogen source—proteins,
	        amino acids,
	        ammonium
	        cpds.
	Mineral source
	Access. gwth. subs.
	
	Water
		Carbon source—can use starch
		Nitrogen source—
		
		Mineral source
		Access. gwth. subs.—can synth.
		        most
		Water

Source	Host plant—potato or tomato Obligate parasite	Dead organic remains (give examples)

Absorption mech.	Haustoria from intercellular
	   hyphae. Drawing.
	Digestive enzymes
	Death of host.
		Finely branched hyphae ramify-
		   ing through substrate. Diges-
		   tive enzymes. Decay

Reprod. Methods	Asex. and sex	Asex. and sex

Asex. form and structures	Spgia.—aerial hyphae
	when wet—zoospores
	dry—conidia
	   Drawing
		Spgia.—erect hyphae
		Sporangiospores only
		   Drawing
		Sprout cells

Sex. form and structures	Oogamy, homothallic
	Oospore
	Where meiosis
	   Drawing
		Isogamy, heterothallic
		Conjugation of gametangia
		Zygospore—sporangium
		Where meiosis
		   Drawing

Answer

The selected fungi are the parasitic *Phytophthora infestans* and the saprophytic *Mucor mucedo*.

They are nutritionally comparable in that they both lack the ability to use light or other energy source with which to synthesize their food requirements from simple inorganic nutrients. They are said to be heterotrophic and obtain energy and materials from organic compounds previously synthesized by autotrophic forms.

Their nutritional requirements are broadly similar also. Both require sources of carbon, nitrogen, minerals, accessory growth substances and water. There are some similarities and differences between the two with regard to the exact nature of these sources. Whereas *M. mucedo* can make use of starch as a carbon source, *P. infestans* cannot and is presumed to use a simpler carbohydrate such as glucose. Both can utilize proteins, amino acids or ammonium compounds as sources of nitrogen and both require phosphates and sulphates. *M. mucedo* is known also to require potassium, magnesium and iron; it is presumed that the parasite will have similar requirements. *P. infestans* is known to require a wide variety of accessory growth substances such as riboflavin, thiamin, biotin and pantothenic acid whereas the saprophyte is able to synthesize most of these.

The major nutritional difference between these plants lies not in the mode of nutrition, nor fundamentally in the nutritional requirements, but in the sources from which these requirements are taken. Whereas *M. mucedo* is able to obtain its food substances from dead organic remains such as bread and other vegetable remains by digesting them through the agency of extra-cellular enzymes, *P. infestans* finds its requirements only in the living bodies of either the potato or the tomato which it infests as an obligate parasite. Coupled with this difference in modes of life are seen differences in mycelial structure. In *P. infestans* specialized absorbing growths, haustoria (*see* diagram), are developed on the intercellular hyphae. These penetrate the host cells and through them the necessary nourishment is extracted. The saprophyte shows no comparable absorbing structures, its hyphae ramify profusely through the substrate without showing specialization of parts.

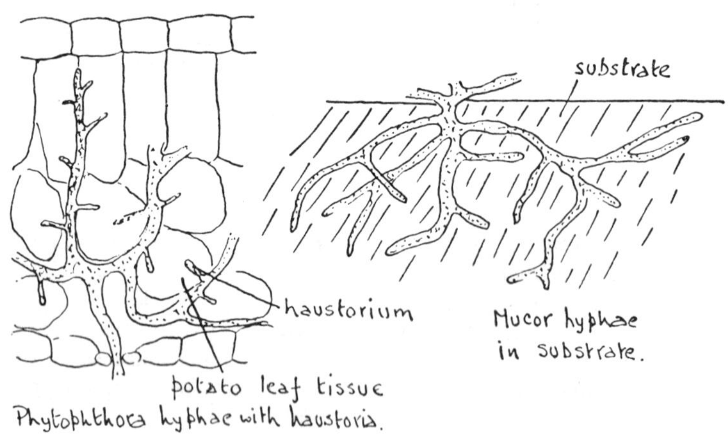

Phytophthora hyphae with haustoria.

In their reproductive processes, the two fungi are again fundamentally similar in that both exhibit asexual and sexual methods of reproduction. There are differences, however, in the ways in which these are effected. *P. infestans* reproduces asexually by developing sporangia on aerial hyphae which grow out from the surface of the host in damp conditions. If damp conditions persist, these sporangia produce many very tiny biflagellate zoospores which upon release swim to new hosts. Under drier conditions, the sporangia are shed whole and act as airborne conidia, infecting new hosts by direct germination. The corresponding asexual structures of the saprophyte are sporangia, borne on erect hyphae above the substrate surface. The sporangia dehisce to release many small brown non-motile, airborne spores. (*See* diagrams.)

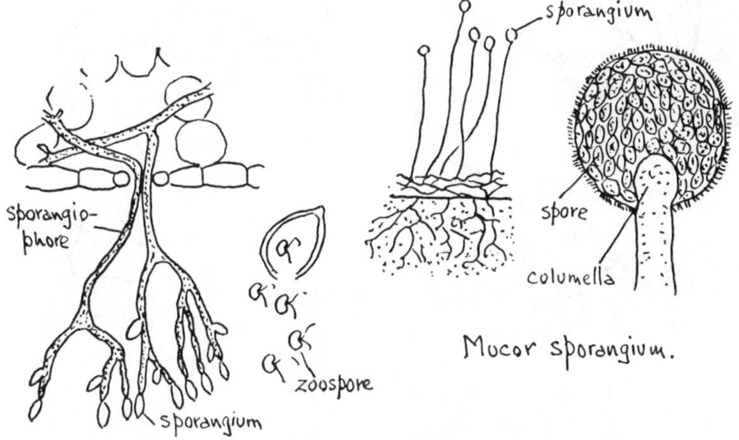

Phytophthora sporangia.

Under certain conditions of growth, *M. mucedo* may develop rounded masses of protoplasm, or sprout cells, within the hyphae. This could be considered to be a form of vegetative propagation since the sprout cells can later give rise to new mycelia. The parasite shows no corresponding method of reproduction.

The sexual process in *P. infestans* can be described as homothallic and oogamous, that is, the male and female gametes may be produced on the same mycelium in antheridia and oogonia respectively (*see* diagrams). The product of gametic fusion is a resting oospore formed inside the oogonium. The corresponding process in the saprophyte differs from this in two main ways. In the first place, no distinctive sex organs are formed, the gametic nuclei being developed in gametangia which are identical in all morphological respects. These are brought together during a conjugation process, fuse as one, allowing the gametic nuclei to mingle and later fuse to form at least one zygote. The product of the conjugation process is a resting zygospore. The second difference to be noted is that the saprophyte is very strongly heterothallic, that is, the fusing gametic nuclei must originate from different strains of mycelium (often known as + and —) or no sexual process can occur.

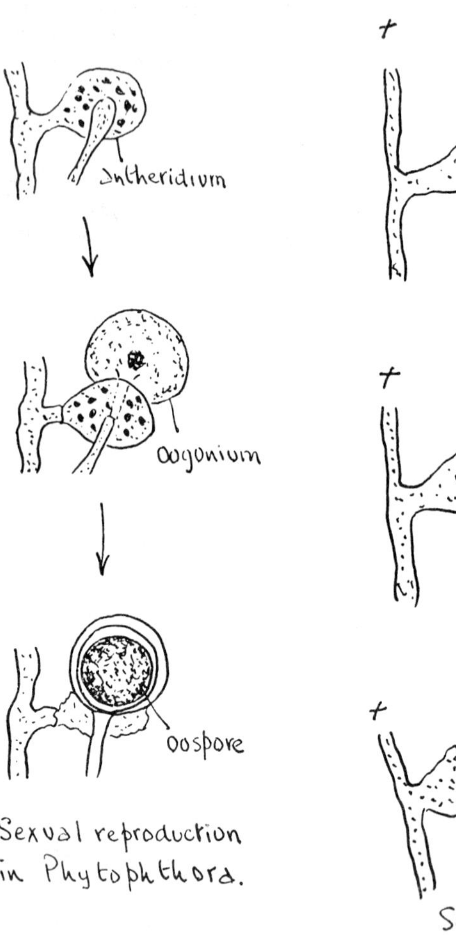

Sexual reproduction in Phytophthora.

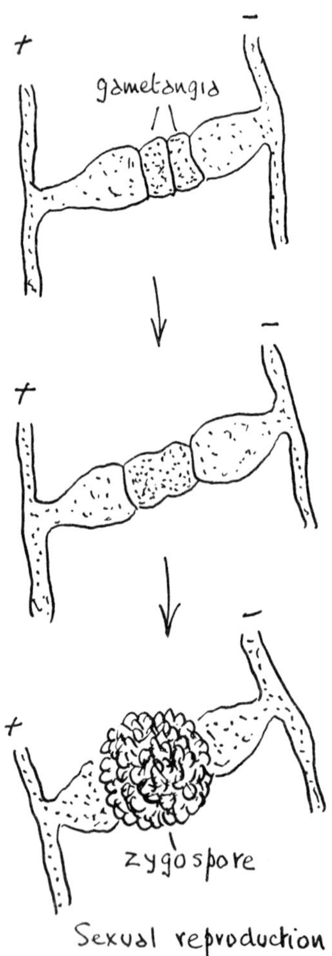

Sexual reproduction in Mucor.

Development of the oospore in *P. infestans* is by germ tube and is preceded by a meiotic division of the zygotic nucleus. In *M. mucedo*, the zygospore content, after meiotic division, develops into an asexual sporangium the spores from which can then develop into new mycelia.

Question

By reference to three *named* plants belonging to different divisions of the plant kingdom, explain what is meant by "alternation of generations."

Plan

State plants selected and divisions.
Give brief statement on meaning of a.of g.
Draw life cycle diagrams of three plants. Point out common
   sequence.

Refer to *Fucus* to show no a.of g., i.e. one body form only, developed directly from zygote, therefore diploid, this always produces next generation directly by a sexual process. Meiosis immediately preceding gamete formation. No haploid form produced.
Refer to *Pellia* to show a.of g. i.e., two body forms, one directly from zygote, therefore diploid (sporophyte); this reproduces asexually after meiosis to give second haploid body form (gametophyte); this reproduces sexually to produce the diploid sporophyte once more.
Refer to *Dryopteris* to show essential similarity to *Pellia*. Point out differences. Drawings.

Answer

The plants chosen for reference are *Fucus vesiculosus* (*Phaeophyta* *Pellia epiphylla* (*Bryophyta*) and *Dryopteris filix-mas* (*Pteridophyta*). They will be referred to by their generic names only hereafter.
Alternation of generations refers to the condition in some plants in which the continuity of the line of descent of the species is maintained through generations of haploid and diploid body forms alternating with one another, the haploid producing the diploid by a sexual process and the diploid giving rise to the haploid by an asexual process. Examination of the sequences of events in the life cycles of the plants named will show that this condition is not evident in the first but is evident in the other two.

The life cycles of these plants may be represented diagrammatically as follows—

All show a sequence of events common to all sexual cycles. This is:

$$\text{fertilization} \rightarrow \text{zygote} \rightarrow \text{meiosis} \rightarrow \text{gametes} \rightarrow$$

The sequence may be interrupted at one or both of two points by the development of a plant body. The first of these follows directly upon zygote formation by mitosis and the other results from meiotic derivatives of the zygote. When both occur in the cycle, the species is said to exhibit alternation of generations.

In the case of *Fucus*, one body form only appears in the sequence. This is a diploid body resulting from the direct development of the diploid zygote. This, by meiosis gives rise to haploid gametes which fuse to form a diploid zygote once more. The sequence is

fertilization → zygote → diploid *Fucus* thallus → meiosis → gametes →

Fucus does not exhibit alternation of generations, all successive generations being produced from sexually formed zygotes.

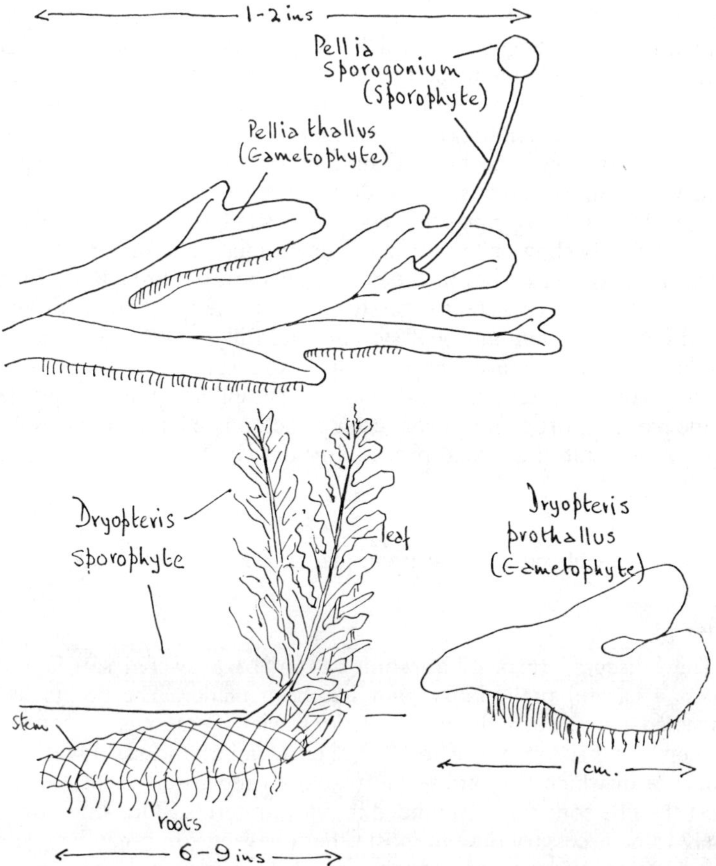

The life cycle of *Pellia* is clearly different. In this plant, the cycle is interrupted at two points by the development of a body form so:

fertilization → zygote → diploid sporogonium → meiosis

→ haploid *Pellia* thallus → gametes →

The diploid sporogonium or sporophyte generation gives rise by meiosis to asexual haploid spores each of which develops into a *Pellia* thallus or gametophyte generation. This, by gamete and zygote formation, gives rise to the next sporophyte generation, and so on, the sporophytes and gametophytes alternating with each other in unbroken regularity. *Pellia* clearly exhibits alternation of generations.

In *Pellia*, the haploid gametophyte is a green thalloid vegetative body which when mature produces distinct male and female sex organs, antheridia and archegonia, in which the gametes are formed. Fertilization occurs inside the female organ. The zygote thus formed, during its development into the diploid sporogonium, becomes physically attached to and physiologically dependent upon the gametophyte. The diploid generation is no more than a stalked spore-producing capsule with very little sterile tissue (*see* drawing on p. 123).

The life cycle of *Dryopteris* is essentially similar to that of *Pellia* and the fern clearly shows alternation of generations. The major difference between the two cases lies in the natures of the alternating body forms. In *Dryopteris*, the persistent vegetative plant is the sporophyte, differentiated into root, stem and leaf systems and fully adapted to terrestrial conditions. It produces haploid spores asexually, each of which develops into a small, green, thalloid gametophyte quite independent of the parent sporophyte. From each gametophyte another sporophyte arises as a result of a sexual process (*see* drawing).

Question

Discuss the effects of wind upon plants.

(*Note*—

The "discuss" form of question cannot be answered satisfactorily unless a careful preliminary plan has been made. The points to be discussed about such things as theories, views, statements, concepts, incidents and so on, must first be called to mind and set down in the sequence in which they are to be treated. The various aspects of each must then be concisely expounded in relevant terms. Drawings are less likely to be necessary than in most other questions, but may be made if by so doing many words can be saved. No personal opinion concerning any point of discussion need be given unless this is expressly demanded. Do not waste time with meaningless introductory and concluding paragraphs.

In this particular question, note that the reference is to "plants." The answer must therefore apply to plants in general, not just one plant or group of plants.)

Plan

Under which environmental conditions wind effects chiefly felt—direct or indirect.

Kinds of effects—
drying, cooling
dispersal
distortion and damage
mixing of atmospheric constituents.

Nature of effects—

for each of above give consequences—beneficial or adverse—adaptations to heighten former or reduce latter.

Overall assessment—

plant life in a motionless atmosphere?

Answer

The direct effect of an air current can be felt only by those plants in its path, thus only terrestrial plants can come under the influence of wind in this way. Aquatic plants may be indirectly affected. Some of the effects are beneficial whilst others are distinctly adverse.

The chief direct effects of wind on land plants are of four main kinds, namely, a drying effect, with a subsidiary cooling effect, an effect on the scattering or dispersal of reproductive parts, that of distortion or damage to parts and the mixing of atmospheric constituents.

The drying effect of the wind is one experienced by all land plants at some time or another and is due to the increased rate of evaporation of water from exposed surfaces, caused by the rapid removal of water vapour from their vicinity. In the main, the effect is an adverse one since too rapid a loss of water will result in serious interference with metabolism, and if this is prolonged, in death. In the case of the more delicate lower land plants, such as the liverworts and mosses, in which there is frequently little or no protection against desiccation, the effect may be at least partially responsible for the restriction of their occurrence to the less exposed habitats. Among the higher vascular plants which for the most part are to a great degree protected against such rapid transpiration by cuticular or peridermal layers, this effect of wind is less significant. Even in these cases, however, only the xerophytic forms seem able to persist in the most bleak and windswept areas such as heaths and moors. In some areas of the world, where high insolation can lead to dangerously high temperatures in organs such as leaves, the cooling effect caused by increased transpiration due to wind may be considered to be beneficial. For most aquatics, the drying effects of wind do not exist, but in the cases of the inhabitants of ephemeral pools and streams, rapid drying up of these, hastened by the action of wind, must result in the cessation of all activity. Only those plants which can form drought-resistant spores or seeds can survive exposure to desiccation.

The effect of wind upon the water relationships of plants can be considered to be important and far-reaching.

An obviously beneficial effect of wind upon plants is that relating to the dissemination of reproductive parts. Few land plants have evolved not to take advantage of this in some way. In the lower plants, such as the fungi, lichens, liverworts and mosses, and in the pteridophytes, their minute spores can remain airborne for very long periods and be carried great distances on the weakest of air currents. All primitive seed-bearing plants, the gymnosperms, rely exclusively on the wind as an agent of transport of pollen and for the most part have evolved winged seeds. Many flowering plants have evolved on similar lines, being wind pollinated and developing winged, plumed or parachute-like fruits or seeds. For most of these, a perfectly motionless atmosphere could possibly lead to their extinction if only for this reason.

The third effect, of distortion or damage, is normally only experienced by plants subjected to vigorous and prolonged buffeting by very strong winds. The effects of hurricanes are inevitably harmful to nearly all plants in their paths and the larger tree forms are frequently uprooted. The isolated incidences of such wind storms, however, make the significance of their damaging effects very small indeed when the total vegetation of the earth is taken into account. Some interesting distortion effects may be seen on plants growing in constantly windswept positions such as cliff tops, the stunting and bending towards the leeward being due to the destruction of buds on the windward side. A damaging effect caused indirectly by wind is experienced by plants such as sea weeds on the shores of seas and large lakes, where waves, caused by winds, may subject them to vigorous pounding from time to time. Most of these have evolved to withstand such hazards and any damage sustained by individuals is of no significance.

One other effect is of some importance, namely, the rapid mixing effect on the atmosphere's constituents by turbulence. This undoubtedly has its benefits to plants in keeping the air in a more or less constant composition everywhere, so that the important gases, carbon dioxide and oxygen, are evenly distributed.

When all these possible effects are collectively considered in relation to the plants which have evolved under, and continue to survive in, conditions where air currents are a permanent feature of the environment, it is clear that the beneficial effects more than balance the adverse. It would be an interesting exercise to ponder the nature of the vegetation of a planet in which the atmosphere was perfectly and permanently still.

Question

Write an essay on the mammalian kidney.

Plan

Morphology, location, general relationships. *Diagram.*
Blood supply and innervation.
Anatomy. *Diagram.* L.S. Kidney.
Histology. *Diagram* of a tubule. *Histol.* Bowman's capsule.
Functions. *Functional diagram* of tubule.
Embryonic origin.

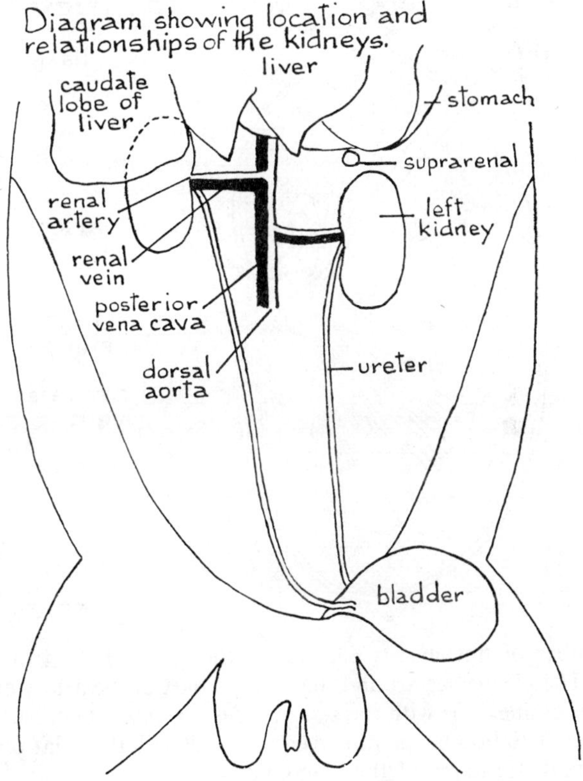

Answer

The kidneys are dark red, bean-shaped structures situated dorsally above the splanchnocoel on either side of the lumbar vertebrae. The right kidney is more anterior than the left (in the rabbit), and in ventral view, it is almost obscured by the caudate lobe of the liver. From the mid-point of the median side of each kidney, a white ureter passes posteriorly to the ventral bladder, entering it dorsally (*see* diagram, p. 127).

Each kidney receives a renal artery from the dorsal aorta. In the kidney, the renal artery branches into arcuate arteries and after further branching there are numerous specialized tufts of capillaries. The blood from these is eventually collected into arcuate veins which empty into the renal vein. The renal vein leads to the posterior vena cava in which the blood is returned to the heart (*see* diagram).

A number of fine nerves pass along the renal artery from a renal plexus. The plexus lies around the median part of the artery and it has many fine connexions with the solar plexus. The whole nerve supply is autonomic with both sympathetic connexions with the solar plexus and parasympathetic fibres of the vagus nerve.

Each kidney is enclosed in a tough connective tissue capsule. Within the capsule there are three distinct regions; an outer cortex with a granular appearance, an inner medullary zone which is striated and at the innermost region there is a small cavity called the pelvis. This cavity is formed by the expanded end of the ureter. The medullary zone bulges as a pyramid with a small pore leading into the pelvis (*see* diagram).

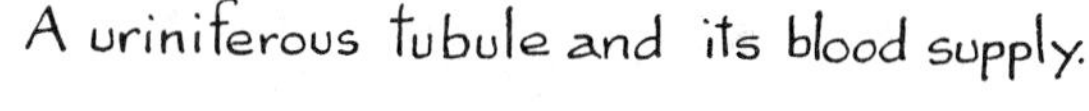

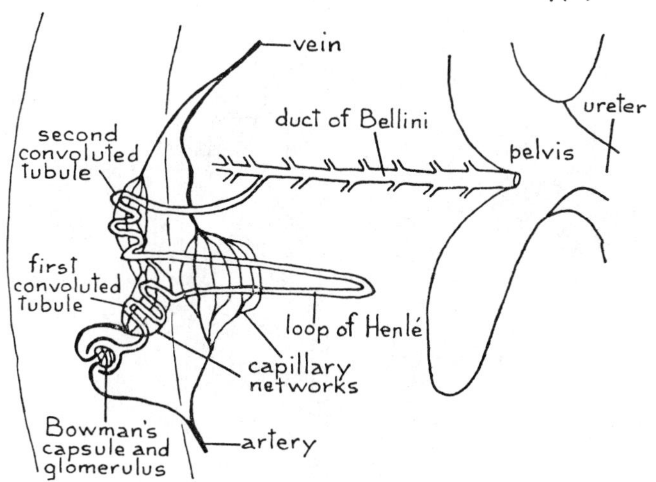

Apart from connective tissue and blood-vessels each kidney contains a very large number of uriniferous tubules (*see* diagram). Each tubule ends blindly as a small rounded funnel in the cortex; the funnels are known as Bowman's capsules, each consisting of a double layer of squamous epithelium enclosing a small cavity. From each Bowman's capsule a convoluted tubule with cubical epithelium winds about the cortex and from it the descending limb of Henlé passes into the medulla. It turns back at the loop of Henlé, and the ascending limb of Henlé re-enters the cortex, where it continues as the second convoluted tubule. This, with many similar tubules, enters a straight collecting tubule which proceeds through the medulla to open at the apex of the pyramid in the duct of Bellini.

Entering the narrow opening of each Bowman's capsule is a small arteriole. In the capsule, it breaks up into a network of capillaries, the glomerulus, and the blood from these is collected into an afferent arteriole (*see* diagram). This leaves the capsule and soon sub-divides to give a rich network of capillaries enveloping the convoluted tubules. The blood from these is collected into a venule, and after passing into larger veins, it leaves the kidney by the renal vein.

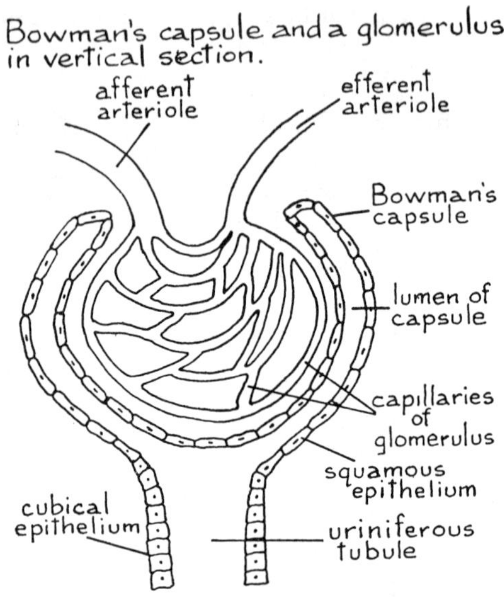

The kidney has two functions of equal importance; they are osmoregulation and excretion. Owing to the fact that the efferent arteriole of each glomerulus is smaller than the afferent arteriole, a high blood pressure is developed. Under this pressure, ultrafiltration takes place from the glomerular capillaries through the thin epithelium of Bowman's capsule into the lumen of the tubule. The filtrate consists of plasma of the blood together with all the substances in true solution; colloids such as blood proteins do not pass from the blood. Along the tubule, especially in the convoluted portions, water, glucose, various salt ions and other useful substances are resorbed. The whole process can be summarized as active and selective resorption so that unwanted excretory products such as urea are passed out of the body in the urine, while the water and mineral salt content of the blood are carefully regulated, excess of either being eliminated (*see* diagram, p. 131). On average, the urine consists of 95 per cent of water, 2 per cent of urea and 3 per cent of other materials.

The mammalian kidney is a metanephric kidney, the pronephros and mesonephros having degenerated. It is developed partly from the nephrotomes of the lumbar region and partly from a backgrowth of the posterior region of the mesonephric (Wolffian) duct.

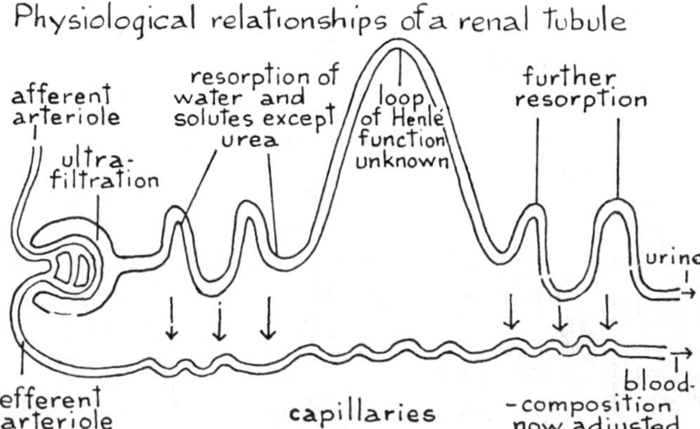

By means of annotated diagrams, outline the life-history of the malaria parasite. Show how a knowledge of this life-history has been of value in reducing the incidence of malaria.

Plan of Answer

Intro. statement—3 species—the fevers—the hosts.
Life cycle in man—infection—schizogony in liver—schizogony in corpuscles—gamogony.
Life cycle in mosquito—infection—gametocytes—gametes—zygote —encysted—sporozoite formation—bursting of cyst—salivary glands— re-infection.
Vulnerable points in cycle of parasite and mosquito—measures taken—success.

Answer

"Malaria parasite" refers to a number of species of the genus *Plasmodium* which cause malaria in man. The three common species are *P. vivax* (tertian fever), *P. malariae* (quartan fever), *P. falciparum* (continuous fever). All have similar life-history. All have two hosts, a mammal and an insect vector. *P. vivax* is described here; the hosts are man and the female mosquito *Anopheles maculipennis*.

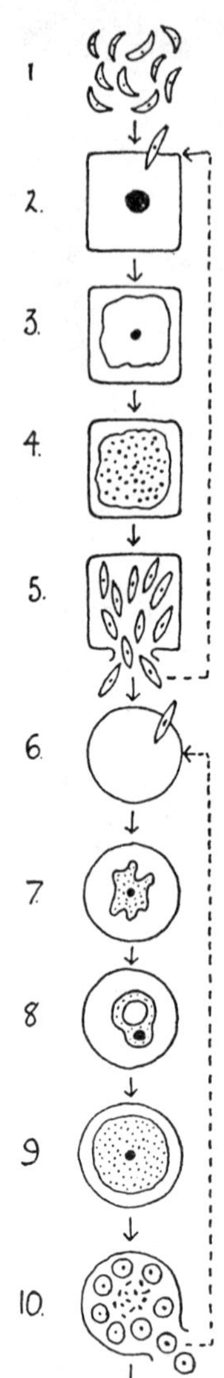

1. Mosquito pierces skin: injects saliva containing sporozoites, 2 μ long. After one hour all have left blood and are in the liver.

2. Each sporozoite penetrates liver cell—consumes contents.

3. Growth to form a large spherical schizont—takes about 7 days.

4. Repeated mitosis—over 1,000 nuclei formed: partition of cytoplasm.

5. Schizozoites released into blood, 3–4 μ long—some re-enter liver cells—reservoir of infection—most enter red corpuscles.

6.

7. Feed and grow: become amoeboid.

8. Large vacuole appears: signet-ring stage.

9. Nearly fills corpuscle.

10. Division to form about 30 rounded schizozites. Released—residual excretory granules cause fever. Infect further corpuscles.

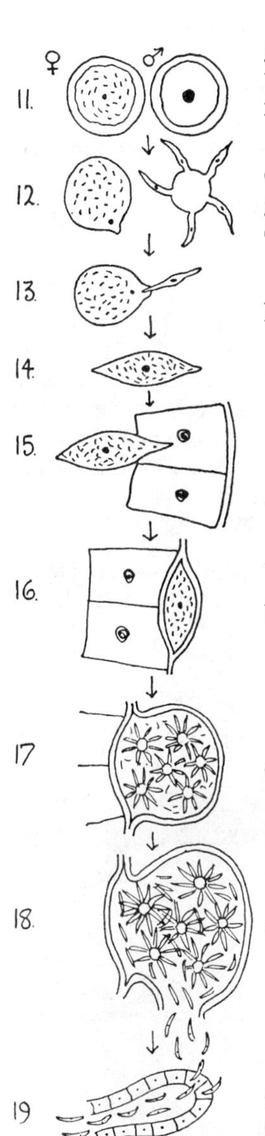

Eventually growth—no division—form gametocytes—male with large nucleus—female with small nucleus and yolk granules. No further development in man.

Ingested by mosquito; in crop all stages digested except gametocytes which form gametes. Female one large gamete; male 8 elongated.

Fertilization—male gamete enters at small papilla.

Motile spindle-shaped zygote formed.

Zygote penetrates wall of crop.

Encystment and growth in connective tissue bordering haemocoel.

Very large cyst; sporozoites formed by division around borders of small vacuoles.

Cyst bursts; sporozoites in haemocoel—swim to salivary glands.

Sporozoites in salivary glands—now infective. No further development till injected into man with saliva.

Knowledge of the life-cycle discloses the most convenient points at which to destroy the parasites or to prevent infection. When the habits and life-cycle of the vector are known as well, great progress can be made.

We know that mosquitoes are nocturnal, cannot fly far, and rest under vegetation in the daytime. They lay their eggs in water and have aquatic larvae and pupae which must breathe at the surface. Prophylactic measures based on knowledge of these habits and the life-cycle. Prevention of mosquito-bites will prevent malaria. Hence netting of beds and windows, avoidance of travel by night, clearing of vegetation around villages, covering of water containers, spraying paraffin on large bodies of water, drainage of marshes, are all widely practised and are very effective. Spraying huts with solution of D.D.T. in kerosene kills adult mosquitoes resting in the houses.

To prevent disease, good modern drugs are available: mepacrin, atebrin and paludrin. These kill the sporozoites in the blood before they reach the liver, and also kill the schizozoites in the blood. Nevertheless, a reservoir of infection in the liver may remain for years.

By combination of all methods, notable success has been achieved, e.g. in Cyprus and British Guiana the disease has been eliminated.

Question

Write brief illustrated notes on five of the following—solar plexus, Eustachian tube, lateral line organ, trophoblast, synovial joint, recurrent laryngeal nerve.

Plan

(For each, adopt the same plan.)

Where found? What groups of animals and what location.
Diagram showing structure and relationships.
Functions.
Any special significance, especially evolutionary.

Answer

(*N.B.* All six are done here.)

The solar plexus. Occurs in all vertebrates. Consists of coeliac and anterior mesenteric collateral ganglia. Found in mesentery dorsal to stomach around anterior mesenteric artery. Nervous connexions with sympathetic ganglia and structures served are shown in diagram (for mammal, p. 135).

General effects of this sympathetic outflow are to constrict blood-vessels, inhibit peristalsis and secretion except for adrenalin and to stimulate breakdown of glycogen in liver. Preparation to meet shock; allied with hormonal effects of adrenalin. Opposite effects by para-sympathetic fibres of the vagus.

Relationships of the solar plexus.

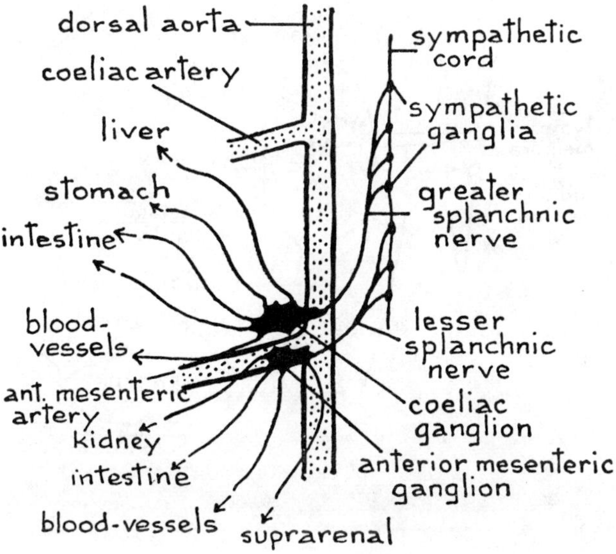

Eustachian Tube. Occurs in amphibians, reptiles, birds and mammals. Paired tubes from pharynx to tympanic chamber which has ear-drum externally and fenestrae ovalis and rotunda internally. Function to allow air to pass from pharynx to tympanic chamber to equate internal air pressure with external atmospheric pressure. Valvular arrangement at pharyngeal end—opened to equate pressures when swallowing. In evolution, probably derived from spiracular cleft; the only cleft to persist in adult terrestrial vertebrates.

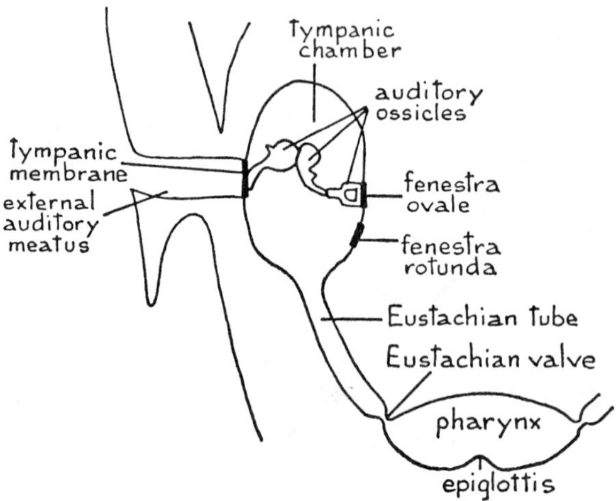

Lateral Line Organs. Occur in lateral line system of fishes and tadpole stages of amphibians. System consists of network of canals over head and two lateral lines to tail. (*See* diagram, p. 137.) Canals filled with jelly; open to surface at intervals. Contain neuromast organs— each has small group of sensory cells—cytoplasmic tips project into jelly—nerve fibres pass from bases. Function is perception of mass movement or vibration in the jelly—transmitted from the external medium down the openings. Value—shoaling, detection of moving or stationary obstacles. Important sense in fishes—compensates for poor eyesight. In evolution, the ear probably has origin as lateral line organs—sense organs in ear are of same nature.

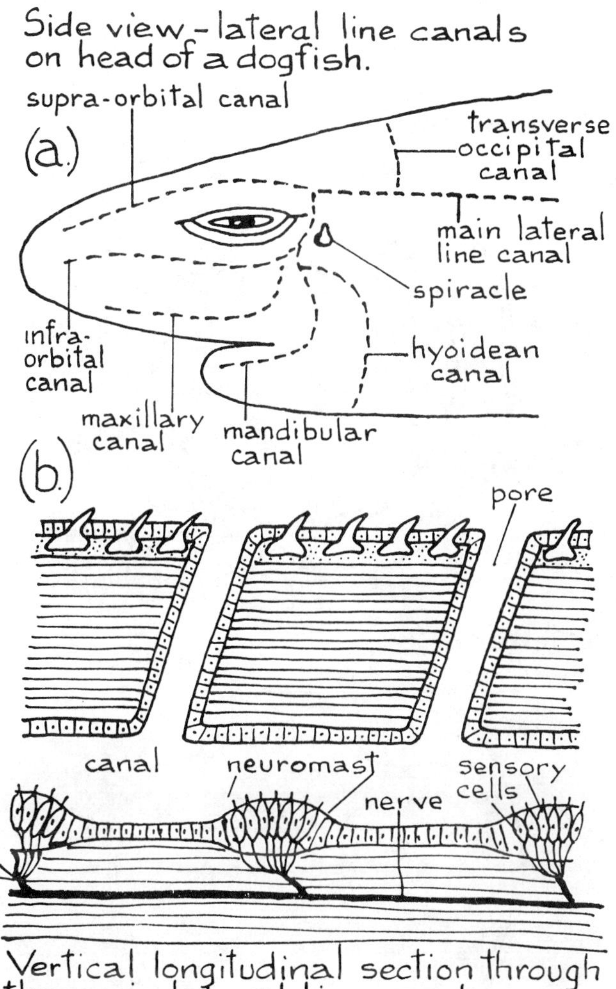
Side view - lateral line canals
on head of a dogfish.
supra-orbital canal
(a.)
transverse
occipital
canal
main lateral
line canal
spiracle
hyoidean
canal
infra-
orbital
canal
maxillary
canal
mandibular
canal
(b.)
pore
canal
neuromast
nerve
sensory
cells
Vertical longitudinal section through
the main lateral line canal

The Trophoblast. Occurs in mammalian embryos. After morula stage, cavity separates outer layer from inner mass. Fluid in cavity, nutritive secretion of uterine glands. Outer layer is trophoblast; inner mass is embryonic knob from which all the embryo develops.

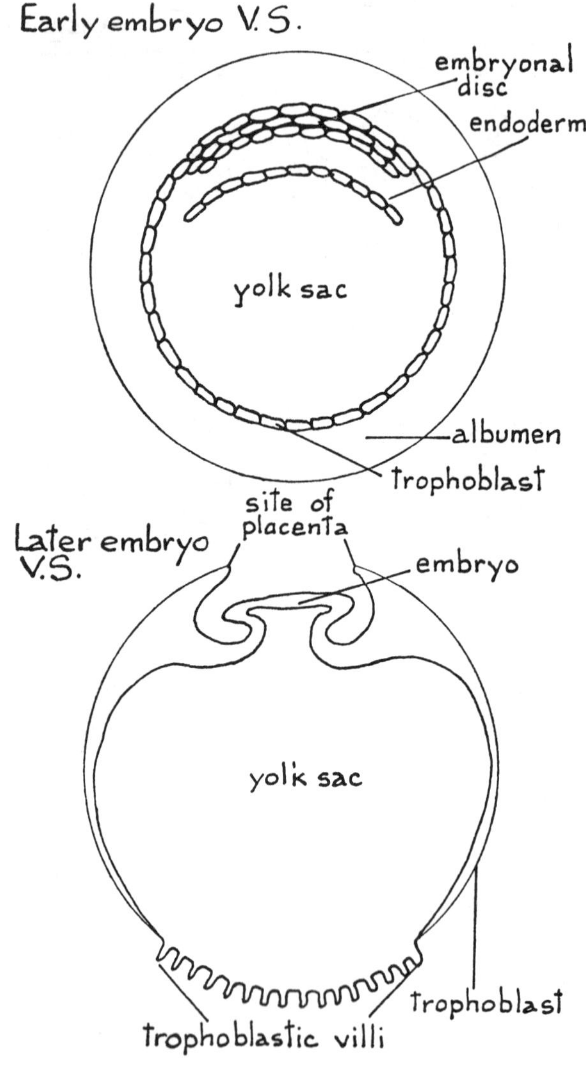

Trophoblast opposite embryo produces villi—grow into depressions of uterine wall—absorb nutritive secretion. Later these villi degenerate and placenta formed diametrically opposite—trophoblast plays important part in formation of placenta. Functions of trophoblast—temporary implantation and nutrition—later helps in permanent implantation.

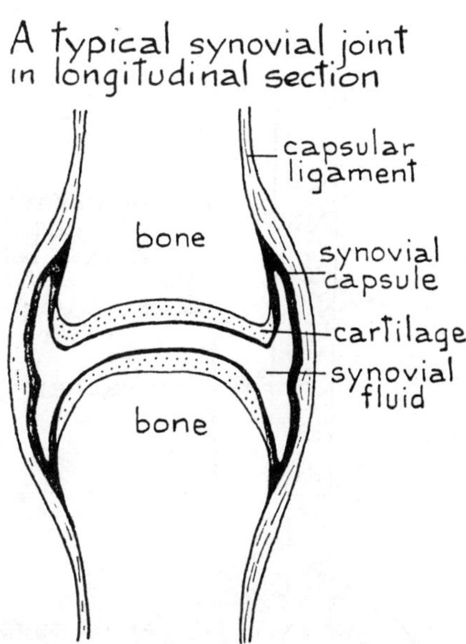

Synovial Joint. Occurs in terrestrial vertebrates at all joints where there is free movement. Surrounding joint is a capsular ligament like a sleeve fitted over ends of both bones. Between bones is a connective tissue capsule containing synovial fluid. Cavity may be subdivided by cartilages. Fluid reduces friction and absorbs shock. (Diagram shows condition for mammal.)

Recurrent Laryngeal Nerve. Occurs in reptiles, birds, mammals—one of corollaries of evolution of a neck. Larynx supplied by two branches of vagus—anterior laryngeal and posterior laryngeal—

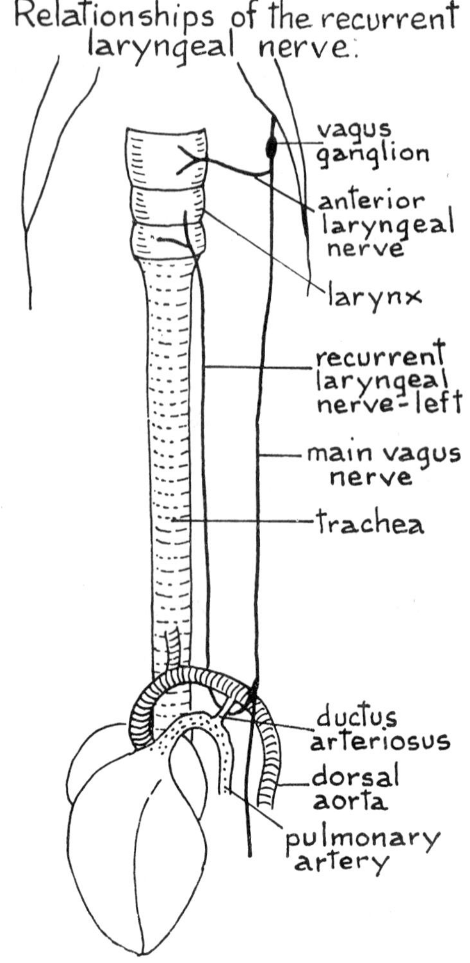

originally. With interpolation of extra vertebrae these two nerves widely separated, posterior leaves vagus at base of neck, doubles forward close to trachea to reach posterior part of larynx. On right, the nerve loops around subclavian artery; on left, round the ductus arteriosus.

APPENDIX

ANSWERS TO QUESTIONS

The Meaning of "Being Alive" (p. 8)

1. Presumably to disintegrate into the highest state of uniform disorder in which all usable energy has been made non-available or the free energy is at a minimum. 2. It possesses the ability to reverse this tendency towards maximum entropy and it can also build more molecular structures of its own kind. 3. To perform the various kinds of work that have to be done to maintain and re-produce the system. 4. To repair, replace or add to existing structural com-ponents and for making the components of a new generation of such systems. 5. (a) the sun, (b) the inorganic environment. 6. By replicating itself. 7. When it can no longer extract and make use of energy and materials from its sur-roundings. 8. It tends to disintegrate in the same way as any non-living system.

Cell Structure and the Nature of Protoplasm (p. 9)

1. The cellular condition exists when a whole protoplasmic system (organism) is composed of more or less discrete units of protoplasm separated from one another by at least their surface membranes or more often by inert inter-cellular substance. There may be some continuity between individual cells by way of fine intercellular connections (plasmodesmata). Most plants and animals are cellularized. A protoplasmic system functioning as an organism is said to be non-cellular when it is not separable into such units. Slime fungi, protozoa and some other microbes are of this form. 2. A protoplasmic system not separable into cells, multinucleate and derived by repeated replication of the original nucleus and addition of cytoplasm without cleavage of this. Some fungi and algae are coenocytic and sarcoplasm of striped muscle is said to be syncytial, an equivalent term used in zoology. 3. A protoplasmic unit including any external surface covering secreted by it. 4. A surface membrane enclosing the cytoplasm and nucleus. 5, 6. Vol. I, Chap. 1. The Living Cell of a Plant. 7. By the intercellular material of the middle lamella. 8. Most commonly of micro-fibrillar cellulose in a matrix of hemicellulose and pectic compounds which may be added to in specific instances by deposits of cutin, suberin or lignin. Chitin may replace cellulose in the cell walls of some fungi. 9. (a) Pits originate at the early stages of wall formation at zones where wall material between cells is not secreted. (b) Intercellular spaces most commonly arise as the walls of enlarging cells pull away from one another (schizogenous origin). They can sometimes form when some cells in a group disintegrate to leave a space (lysigenous cavity). In some instances, e.g. aquatic plants, spaces be-tween cells are present from earliest cell formation at a meristem (ontogenous origin). 10. Connecting adjacent plant protoplasts most commonly through pitted areas. 11. Vol. I, Chap. 1. The Origin and Development of the Mature Cell. 12. The cell increases in size and often changes shape, vacuolates intern-ally and the wall matures through primary to secondary layer formation. In

some cases when the wall is fully developed the protoplast may disappear leaving an empty cavity. 13. Vol. I, Chap. 1. The Living Cell of an Animal. 14. Living components take part in the processes essential to the maintenance of the system in its living state. Non-living components are inanimate products of the activity of the living substance. They are sometimes called ergastic substances. 15. Mitochondria; ribosomes; Golgi bodies. Vol. I, Chap. 1 and Vol. II, Chap. 2 (under these names). 16. (a) Chloroplasts. (b) Lysosomes. Vol. I, Chap. 1 and Vol. II, Chap. 2. 17. Vol. I, Chap. 1. Non-living Inclusions of the Cytoplasm. 18. Nucleoplasm, nucleoli and chromosomes enclosed within the nuclear membrane. 19. Nucleoplasm is a fluid matrix containing DNA and protein; nucleoli are composed chiefly of RNA and chromosomes of DNA. 20. Nucleoplasm forms a liquid phase for the other component activities; nucleoli are possibly sites of RNA formation; chromosomes are carriers of inheritance information coded in DNA. 21, 22. Vol. I, Chap. 1. Differences Between Plant and Animal Cells. 23. Surface area/volume ratio and cytoplasm/nucleus ratio. 24. Vol. II, Chaps. 1 and 2. 25. The free energy of a living system tends to increase rather than decrease when considered as an isolated system and it can replicate its own structure. 26. Water 85–90 per cent; protein up to 10 per cent; carbohydrate up to 2 per cent; fats up to 2 per cent; inorganic salts 1–2 per cent. 27. Protoplasm varies in detailed composition from individual to individual and from time to time in the same individual according to the precise nature and rate of its activity. 28. A heterogeneous colloidal dispersion system in a continuous aqueous phase. 29–42. Vol. II, Appendix. Properties of Substances in the Colloidal State. 43. Radiant energy (light); chemical bond energy. 44. Chemical bond energy of ATP. 45. Heat production; executing mechanical and chemical work; maintenance of the protoplasmic system as an ordered system of molecules; osmotic work; maintenance of electric potentials across membranes; transmission of electrical impulses and sound and light emission in some cases. 46. As chemical bonds in either ATP (plants and animals), soluble, transportable sustances such as glucose (plants and animals) or insoluble non-motile "storage" compounds such as starch (plants), oils or fats (plants and animals). Glycogen is an equivalent to starch for animals but it is not secreted as solid grains.

Cell Replication (p. 10)

1. From an original parent cell that forms the starting point for another generation. 2. By replication of the cell substance and division into separate cells. 3. Nuclear division, cytoplasmic cleavage, cell separation. 4. Vol. I, Chap. 2. Mitotic Division. 5. They are not significantly different, but the origin of the spindle may vary. 6. In both, the chromosomes replicate as chromatids and then separate into two equivalent sets of genetic material. 7. As an aid to identifying stages in the process. Prophase-chromosomes appearing; metaphase-chromosomes lying at equator of spindle; anaphase-chromatids moving to spindle poles; telophase-chromatids becoming extended as in the interphase state. 8, 9. Vol. I, Chap. 2. Cell Separation. 10. Cellulose arranged as micro-fibrils in an amorphous matrix of hemicellulose and pectic substances. 11. A further deposition of wall substance internally

on the primary layer, secreted later in the cell differentiation process, composed of cellulose with or without other components such as lignin, cutin, suberin. 12. Vol. I, Chap. 2. Cytoplasmic Cleavage; Cell Separation. 13. Meiosis occurs for most animals immediately preceding gamete formation (gametogenesis) and for most plants immediately before spore formation by the sporophyte generation (see life cycle diagrams in Vol. I). 14. Vol. I, Chap. 2. Meiotic Division of Cells. 15. First prophase (leptotene, zygotene, pachytene, diplotene, diakinesis); first metaphase; first anaphase; first telophase (may be omitted); second prophase (may be omitted); second metaphase; second anaphase; second telophase. 16, 17, 18, 19. Vol. I, Chap. 2.

Material Exchanges between Cells and External Solutions (p. 12)

1. The cell membrane (plasma membrane, plasmalemma). The inner surface layer at the vacuole of a plant cell is called the tonoplast. 2. Protein-lipide layer acting as a diffusion barrier, common to both plant and animal cells; if destroyed the internal cell substance is dispersed into the external fluid; in effect, the cell membrane isolates the living structure from the inorganic surroundings. 3. By an aqueous phase continuous through the protoplasm and cell membrane and beyond. 4. Exchanges of materials. 5. Immerse it in strong sugar solution (or equivalent hypertonic solution) and then place it in pure water, noting changes in size of the protoplast. 6. Plasmolysed. 7. Deplasmolyse by placing in pure water. 8. Immerse in saline solutions of varying strengths. 9. The concentration of dissolved substances, i.e. the number of dissolved particles per unit volume; the molar concentration of solutes. 10. Solutes are said to diffuse, solvents to osmose. 11. No. (a) All particles pass unimpeded; (b) no particles pass; (c) solvent particles only pass. 12, 13. Vol. II, Chap. 3. Osmosis. 14. The volume of the weaker decreases; that of the stronger increases. 15. Solvent flows from the weaker to the stronger solution. 16, 17, 18, 19, 20. Vol. II, Chap. 3. Osmosis; The Physical Laws of Osmosis. 21. 1 atm (101 325 N m^{-2}) at 0°C. 22. 22·4 atm (22·4 × 101 325 N m^{-2}) at 0°C; in fact, nearer 25·0 atm. 23. Given by the molar concentration times 22·4; 100/342 × 1000/500 × 22·4 = 13·1 atm. 24. Its dissociation into ions. 25. 10·1/101 × 1000/100 × 2/1 (dissociation factor) = 44·8 atm. 26. By summation of individual osmotic pressures, in accordance with the total concentration of all the solute particles of all kinds. 27. Their volumes shrink because they lose water to the external solution. 28. In the above case, none. 29. Their volumes tend to increase because they may gain water from the external solution. 30. It prevents it from swelling to rupture point as the internal hydrostatic pressure rises; in effect it limits the intake of water by the protoplast. 31. Osmotic pressure of the external solution (OP_e) tending to cause water to move out; osmotic pressure of the internal solution tending to cause water to move in (OP_i); wall or turgor pressure exerted by the cell wall tending to resist entry of water as the cell becomes more turgid (TP). See also Vol. II, Chap. 3. Water Movement in a Biological System—Elementary Thermodynamical Treatment. 32. Diffusion pressure deficit (DPD); previously called suction pressure. 33. $DPD = (OP_i - OP_e) - TP$. 34. The cell is tending to lose water. 35. (a) Turgid; (b) flaccid. 36. Vol. II, Chap. 3.

Plasmolysis and Related Phenomena. 37. They become fully turgid if plant cells; they tend to swell and rupture in animals. 38. They must eliminate any excess water taken in, i.e. osmoregulate. 39. The cell wall prevents intake of excess water. 40. It can readily replace any water lost by evaporation and can thus maintain full turgidity. 41–53. Vol. II, Chap. 3. Exchange of Dissolved Substances between Cells and their Surroundings.

Hydrogen Ion Concentration and Buffering (p. 14)

1. The proportion of free H^+ to OH^- ions. An acid solution contains excess H^+; an alkaline solution, excess OH^-. 2. A solution containing equal numbers of H^+ and OH^- ions. 3. Its full potential for releasing H^+ ions into solution. 4. Both require the same amount of alkali to neutralize them, but hydrochloric acid much more readily dissociates into its constituent ions. 5. Degree of acidity. 6. Because it contains equal quantities of H^+ and OH^- ions. 7. 10^{-14}. 8. 10^{-7}; 10^{-7} (hence the product is 10^{-14}). 9. Because changes in one are exactly balanced by changes in the other. 10. As the pH or the negative logarithm of the hydrogen ion concentration. 11, 12, 13. Vol. II, Chap. 3. The meaning of pH. 14. By indicators for colourless solutions and by an electrical method (pH meter) for others. 15. Solutions containing buffers can resist violent changes in pH. 16. (*a*) tend to be constant in sea-water and animal body fluids (*b*) tend to be variable in soil solutions and fresh-water. 17. Mammalian blood, pH 7·4. 18. Animals little; plants commonly much more. 19. Calcifuge; (acid lovers) any of *Rhododendron* spp.; *Calluna vulgaris*; *Erica cinerea*s *Rumex acetosella*. Calcicoles (alkali lovers) any of *Leontodon hispidus*; *Poterium sanguisorba*; *Bromus erectus*. 20, 21. Vol. II, Chap. 3. Some Effects of pH changes on Living Things.

Energy Exchanges (p. 15)

1. Both utilize energy in the performance of work. 2. The energy of chemica bonds (in ATP). 3. The sun (light energy). 4. Photosynthesizers—plants with the necessary energy-capturing substances (chlorophylls and others). 5. From the chemical bond energy of substances absorbed by them. 6. As heat; light; sound; electrical discharges; in doing mechanical work. 7. To maintain the protoplasm as a functional system (osmotic work); to energize its metabolic activity (chemical work); to energize movements (mechanical work); to make provision for future requirements (stored as potential energy).

The Nature of the Chief Constituents of Protoplasm and the Products of its Metabolism (p. 15)

1. Water. 2. Vol. II, Chap. 7. The Role of Water in the Organism and its Environment. 3. Carbohydrates; proteins; lipides. 4. Vol. II, Chap. 4. 5. It is a derivative of a polyhydroxy alcohol. 6, 7. Vol. II, Chap. 4. The Carbohydrates and their Derivatives. 8. Maltose = glucose (1) + glucose (1); sucrose = glucose (1) + fructose (1); lactose = glusose (1) + galactose (1); cellulose, starch, glycogen = glucose (indefinite numbers); inulin = fructose

(indefinite numbers); stachyose = galactose (2) + glucose (1) + fructose (1); raffinose = glucose (1) + fructose (1) + galactose (1); xylan = xylose (indefinite numbers); araban = arabinose (indefinite numbers). 9. Hydrolysis by mineral acids or by the action of carbohydrase enzymes. 10, 11. Vol. II, Chap. 4. 12. It is the product of a reaction between an alcohol and an aliphatic acid, e.g. ethyl acetate, the product of ethyl alcohol + acetic acid. 13. Fatty acids and glycerol, the latter being common to all simple fats. Examples of fatty acids are oleic, butyric, palmitic, stearic, linoleic and linolenic acids.

14. $CH_2OH.CHOH.CH_2OH + 3\ C_{17}H_{35}.COOH$
$\qquad$ glycerol $\qquad\qquad\qquad\qquad\qquad$ stearic acid

$\qquad \rightleftharpoons CH_2COO.C_{17}H_{35}.CHCOO.C_{17}H_{35}.CH_2COO.C_{17}H_{35} + 3\ H_2O.$
$\qquad\qquad\qquad\qquad$ tristearin $\qquad\qquad\qquad\qquad\qquad\qquad\qquad$ water

15. Oil from seeds of olive, sunflower, coconut, peanut, almond; from rind of orange; leaves of thyme; oil from whale blubber (adipose tissue); lard from pig fat; butter fat from milk. 16. Fats are esters of glycerol, i.e. glycerides; waxes are esters of alcohols higher than glycerol. 17. By hydrolysis with dilute sulphuric acid or by lipase enzymes. 18. Vol. II, Chap. 4. Complex Lipides, Steroids and Sterols. 19. It contains the amino group (NH_2), this having replaced a hydrogen atom in an aliphatic acid, e.g. from $CH_3.COOH$ comes $CH_2.NH_2.COOH$; or from acetic acid comes glycine in this way. 20. Glycine, alanine, phenylalanine, serine, cysteine, histidine, lysine, methionine, leucine, valine, threonine, tyrosine, tryptophane, ornithine, arginine, aspartic acid, glutamic acid; about 15. 21. They have both basic and acidic properties. 22, 23. Vol. II, Chap. 4. The General Characteristics of Proteins.

24. $H_2N.CH_2.CO\overline{|OH + H|}NH.CH_2.COOH$
$\qquad$ glycine $\qquad\qquad\qquad\qquad\qquad\qquad$ glycine

$\qquad\qquad\qquad\qquad \rightarrow H_2N.CH_2.CO.NH.CH_2.COOH + H_2O.$
$\qquad\qquad\qquad\qquad\qquad\qquad$ glycyl-glycine $\qquad\qquad\qquad$ water

25. Peptide linkage. 26. (*a*) Dipeptide; (*b*) polypeptide. 27. Straight chains laterally bonded by S—S links or chains variously folded or spiralled and held by hydrogen bonds. They constitute proteins. 28. By hydrolysis with strong acids or with proteinase enzymes. 29. Metaproteins, proteoses, peptones, peptides. 30. Because they are combinations of varying numbers of amino acids of more than twenty different kinds arranged in a multiplicity of ways. 31, 32. Vol. II, Chap. 4. Classification of Proteins. 33. C, H, O, N, S, P, K, Mg, Ca, Fe, Na, Cl, B, F, Si, Mn, Cu, I. 34. Vol. II, Chap. 4. Elements of Biological Importance. Vol. II, Chap. 10. Nutrition of Green Plants: Nutrition of Animals.

Enzymes and their Role in Metabolism (p. 17)

1. Enzymes are organic compounds (proteins) that catalyse the chemical reactions of metabolism in living systems. They differ from inorganic catalysts (some of which may act correspondingly *in vitro*) by reason of the fact that they are not heat stable (their activity as catalysts is destroyed at temperatures up to

100°C). 2. By adding -ase to its substrate name, e.g. protein -ase; or by reference to the kind of reaction it catalyses, e.g. dehydrogen -ase. 3. The enzyme is restricted in the nature of the reaction it can catalyse. 4. Specific to (1) a particular kind of substrate; (2) a particular chemical linkage. 5. The amount of enzyme present; temperature; pH; Vol. II, Chap. 5. Properties of Enzymes. 6. Vol. II, Chap. 5. Nature of Enzymes and their Mode of Action. 7. Esterases; carbohydrases; proteases; aminases. Vol. II, Chap. 5. Hydrolysing Enzymes. 8. Dehydrogenases; oxidases; peroxidases. Vol. II, Chap. 5. Oxidation-reduction enzymes. 9. Carboxylases; decarboxylases; isomerases. Vol. II, Chap. 5. Miscellaneous Enzymes. 10. Ptyalin—starch to maltose; carbonic anhydrase—carbonic acid into water and carbon dioxide; catalase—hydrogen peroxide into water and oxygen; invertase—sucrose into glucose and fructose; lactic dehydrogenase—converts (oxidizes) lactic acid into pyruvic acid; pepsin—splits proteins into peptones; maltase—maltose to glucose; starch phosphorylase—starch to glucose phosphate; zymase—a group of enzymes converting glucose to ethyl alcohol an carbon dioxide. 11. To show that the catalyst concerned is not heat stable as an inorganic catalyst would be.

The Organism as a Co-ordinated Protoplasmic System (p. 17)

1. A living system can energize and materially sustain itself; grow and develop; eliminate its own unwanted by-products; detect and respond to changes in its surroundings; reproduce; it may be able to locomote (most animals, few plants). 2. Vol. I, Chap. 1. Special Characteristics of Living Organisms; Vol. II, Introduction and Chap. 1. The Living and the Non-living. 3. Vol. I, Chap. 1. Grades of Complexity.

Environmental Conditions (p. 18)

1. The aggregate of the physical, chemical and biotic conditions under which an organism exists. 2. Vol. II, Chap. 6. Nature of Environments. 3. Sea, fresh-water, land. 4. No; in some zones they tend to grade into one another. 5. The inter-tidal zone on a shore; brackish waters at estuaries; lake, pond and river sides. 6. Buoyancy; pressure; movements (tides); temperature; optical properties (light penetration); osmotic pressure. Vol. II, Chap. 6. Physical Characteristics of the Seas. 7. Salinity; dissolved gases; pH. Vol. II, Chap. 6. Chemical Characteristics of the Sea-water. 8. As for 6 and 7 above. 9. The earth's solid crust not covered with water (soil) and the atmosphere. 10. The mineral skeleton; organic material; aqueous solution; soil atmosphere; living population. 11. Vol. II, Chap. 6. The Constituents of Soil. 12. Texture; structure; porosity; nutrient content; water content; pH; temperature; colour; ease of tillage (for man). 13. Vol. II, Chap. 6. Properties of the Soil. 14. Vol. II, Chap. 6. Soil Mineral Skeleton and Soil texture. 15, 16. Vol. II, Chap. 6. Comparison of Sandy and Clayey Soils. 17. Vol. II, Chap. 6. 18. Humidity; nutrient content; temperature; pressure; buoyancy; movements (wind); optical properties. 19. The physical and chemical conditions presented by the living body of another organism or by its

by-products. 20. Differences in water relationships, gaseous relationships and in the mechanical support afforded.

The Classification of Living Things (p. 19)

1. Taxonomy relates to the principles on which a classification (of any kind) is made; systematics refers to the groupings arrived at by the application of the chosen taxonomic principles; nomenclature describes how individuals and groups are named. 2. Kingdom; division or phylum; class; order; family; genus; species. 3. Each species has two names, generic and specific. Linnaeus. 4. The generic name. 5. Vol. I, Chap. 4. Nomenclature. 6, 7, 8, 9. Vol. I, Chap. 4. Taxonomic Methods. 10. Chlorophyta; Phaeophyta; Rhodophyta; Eumycophyta; Bryophyta; Pteridophyta; Spermatophyta. 11. Vol. I, Chap. 4. Classification Tables (Plants). 12. Protozoa; Porifera; Coelenterata; Platyhelminthes; Nematoda; Annelida; Arthropoda; Mollusca; Echinodermata; Chordata and see Vol. I, Chap. 4. Classification Tables (Animals). 13. Vol. I, Chap. 4. Classification Tables. 14. One that gives fullest weight to affinities that indicate evolutionary relationships. 15. Because such evolutionary relationships may never be fully disclosed.

The Range of Living Things—Type Studies (pp. 19–57)

THE PLANT KINGDOM

Reference to the major plant groups and details of examples from each (as listed below) sufficient to answer questions set on pp. 19–37 can be found as follows.

Algal forms. Classification: Vol. I, Chap. 4. A unicellular motile green alga—*Chlamydomonas*; a unicellular non-motile green alga—*Chlorella*: Vol. I, Chap. 5. Colonial motile green algae—*Pandorina, Volvox*; a colonial non-motile green alga—*Pediastrum*; filamentous green algae—*Ulothrix, Oedogonium, Spirogyro*; a coenocytic alga—*Vaucheria*; a diplobiontic green alga—*Ulva*; a haplobiontic brown alga—*Fucus*: Vol. I, Chap. 9.

Fungal forms. Classification: Vol. I, Chap. 4. A saprophytic phycomycete —*Mucor*: Vol. I, Chap. 33. Parasitic phycomycetes—*Phytophthora, Pythium, Peronospora, Albugo* [*Cystopus*]: Vol. I, Chap. 31. Saprophytic ascomycetes —*Aspergillus* [*Eurotium*], *Penicilium, Saccharomyces*; a saprophytic basidiomycete—*Agaricus* [*Psalliota*]: Vol. I, Chap. 33. A parasitic basidiomycete— *Puccinia*: Vol. I, Chap. 31.

Symbiotic algal-fungal forms—various lichens: Vol. I, Chap. 24.

Microbes—bacteria, PPLO, viruses: Vol. I, Chap. 36.

Bryophytes. Classification: Vol. I, Chap. 4. A liverwort—*Pellia*; a moss —*Funaria*: Vol. I, Chap. 11.

Vascular plants. Pteridophytes. Classification: Vol. I, Chap. 4. Homosporous pteridophytes—*Dryopteris, Lycopodium*; a heterosporous pteridophyte—*Selaginella*: Vol. I, Chap. 13.

Spermatophytes. Classification: Vol. I, Chap. 4. Primitive spermatophytes —gymnosperms—*Pinus, Taxus, Cycas*: Vol. I, Chaps 14–15.

Angiosperms. Vegetative morphology: Vol. I, Chap. 17. Histology: Vol. I, Chap. 12. Anatomy: Vol. I, Chap. 17. Floral morphology and anatomy: Vol. I, Chap. 18. Reproduction and life cycle: Vol. II, Chap. 16.

THE ANIMAL KINGDOM

Reference to the major animal groups and details of examples from each (as listed below) sufficient to answer the questions set on pp. 37–57 can be found as follows.

Protozoan animals. Classification: Vol. I, Chap. 4. A green flagellate—*Euglena*: Vol. I, Chap. 7. A parasitic flagellate—*Trypanosoma*: Vol. I, Chap. 32. A free-living rhizopod—*Amoeba*: Vol. I, Chap. 6. A parasitic rhizopod—*Entamoeba*: Vol. I, Chap. 32. A free-living ciliate—*Paramoecium*: Vol. I, Chap. 8. Parasitic sporozoans—*Monocystis, Plasmodium*: Vol. I, Chap. 32.

Parazoan animals. Classification: Vol. I, Chap. 4. Sponges—various. Vol. I, Chap. 10.

Metazoan animals, non-chordate. Classification: Vol. I, Chap. 4. Diploblastic, acoelomate metazoans—Coelenterata—*Hydra, Obelia*: Vol. I, Chap. 10. A free-living triploblastic acoelomate—Platyhelminthes—*Planaria*: Vol. I, Chap. 19. Parasitic triploblastic acoelomates—Platyhelminthes—*Fasciola, Taenia*; Nematoda—*Trichinella*: Vol. I, Chap. 32. Segmented triploblastic coelomates—Annelida—*Lumbricus, Nereis:* Vol. I, Chap. 21. A crustacèan—Arthropoda—Crustacea—*Astacus*: Vol. I, Chap. 22. Free-living insects—Arthropoda—Insecta—*Periplaneta* and others: Vol. I, Chap. 23. Parasitic insects—*Pediculus, Aphis, Anopheles*: Vol. I, Chap. 23. *Pulex*: Vol. I, Chap. 32. A social insect—*Apis*: Vol. I, Chap. 23. A gasteropod mollusc—Mollusca—*Helix*: Vol. 1, Appendix.

Metazoan animals, chordate. Classification: Vol. I, Chap. 4. A primitive chordate—Acrania—*Amphioxus*: Vol. I, Chap. 25. An aquatic craniate—Craniata—Pisces—*Scyliorhinus*: Vol. I, Chap. 26. An amphibian—Amphibia—*Rana*: Vol. I, Chap. 27. A primitive terrestrial craniate—Reptilia—*Lacerta*: Vol. I, Chap. 28. A winged terrestrial craniate—Aves—*Columba*: Vol. I, Chap. 28. Mammals—Mammalia—*Oryctolagus, Rattus*: Vol. I, Chap. 29. Histology of Mammals: Vol. I, Chap. 20. Chordate embryology: Vol. I, Parts of Chaps 25 (Amphioxus), 26 (fish), 27 (frog), 28 (chick), 29 (rabbit).

Water Relationships (p. 57)

1. It is a structural component of protoplasm forming a continuous background phase and it acts as a general metabolite. 2. It is a chemical reagent and provides the reducing hydrogen in photosynthesis; it is a good solvent, allowing ready dissociation of solutes; has a high surface tension; low viscosity; high specific heat capacity; good thermal conductivity; its greatest density is at about 4°C. 3. Vol. II, Chap. 7. Role of Water in the Organism and in its Environment.

Plants

4. Desiccation. 5. They possess adequate water absorbing parts and tend to restrict water loss from their surfaces. 6. Fungi; geophytes. 7. Partially adapted. 8. The soil. 9. Vol. II, Chap. 6. Soil—Water Availability. 10. Rhizoids and roots; the latter. 11. Vol. I, Chap. 17. The Root System and Vol. II, Chap. 7. Water Absorption by the Roots of Higher Plants. 12. The soil surface tensional forces and imbibitional forces. 13. Osmotic forces developed by cells containing solutions hypertonic to soil water. 14. Vol. II, Chap. 7. Water Absorption by the Roots of Higher Plants. 15. It is the loss of water from plant surfaces by evaporation. 16. Because it is simply the condition of a wet surface losing water vapour to an unsaturated atmosphere —a normal physical event. 17. Because they lack means of restricting water loss by evaporation from their bodies; they are easily desiccated (dried out). 18. They are commonly protected externally by a water-proof epidermis or periderm. 19. The pores (stomata and lenticels) in these outer coverings through which water vapour can readily pass when they are open. 20. Virtually water-proof outer layers are perforated to allow essential gaseous exchanges. 21. Leaves. 22. A very large internal surface area (wet) continuous with the dryer atmosphere outside through apertures in the epidermis. 23, 24. Vol. II, Chap. 7. Transpiration from Leaves. 25. It depends on the leaf structure; it can be as high as 99 per cent in sun plants with thick cuticles or as low as 80 per cent in plants less well protected (shade plants); the rest is lost by cuticular transpiration. 26. Vol. II, Chap. 7. Diffusion of Gases through Stomata. 27. It tends to close when the leaf dries out too quickly. 28, 29, 30. Vol. II, Chap. 7. Stomatal Mechanism. 31. Atmospheric humidity; temperature; light conditions; wind; water availability to the plant; atmospheric pressure. 32, 33. Vol. II, Chap. 7. Factors Affecting Transpiration. 34. Xerophytes. 35. Extreme aridness due to very limited rainfall as in deserts, or liquid water not available due to rapid evaporation and drainage as in dunes. 36. The flow of water through a plant resulting from transpiration at its surfaces. 37. Xylem tracheids and vessels. 38. Vol. I, Chap. 12. The Xylem. Vol. II, Chap. 7. Movement of Water through the Plant. 39. Vol. II, Chap. 7. Movement of Water across the Root. 40. Vol. II, Chap. 7. Root Pressure. 41. It could only explain an upward lift to a limited height. 42. Capillarity; a pumping mechanism in the stem; a mechanical pull from above. 43. The last—due to evaporational forces. 44, 45, 46. Vol. II, Chap. 7. The Upward Movement of Water. 47. Evaporation. 48. An aperture or gland at a plant surface through which water is actively expelled or guttated. 49. Vol. II, Chap. 7. Guttation. 50. It may effect an increase in rate of movement of dissolved substances upwards through the plant; it may have a cooling effect. 51. Probably not at all.

Animals

1. Isotonicity refers to the condition of equality of concentration of solute particles in two solutions. A solution is hypotonic to another when it contains fewer solute particles per unit volume. The other solution is then hypertonic

to it. 2. Equated to sea-water in marine coelenterates; annelids and crustaceans excrete hypotonic urine; elasmobranch fishes retain urea in the body fluids; teleost fishes imbibe sea-water and excrete salts via the gills. 3. Vol. I, Chap. 27. Excretion. 4. Vol. I, Chaps. 23 (insects); 28 (reptiles), (birds); 29 (mammals) and Vol. II, Chap. 7. 5. Vol. II, Chap. 7. Intake of Water. Output of Water.

Co-ordination (p. 60)

1. It is the achievement of harmonious functioning of all the body systems. 2. To regulate the activities of the various parts so that they work efficiently for the benefit of the whole organism and to create as near a homeostatic condition internally as possible. 3. By the passage of electrical impulses between parts; the dispersion of soluble substances through body fluids; by physical or mechanical effects.

Plants

4. Hormonal or the dispersal of active substances. 5. They lack the need for rapid perception and response to changing conditions of the environment (internal and external). 6, 7, 8, 9: Vol. II, Chap. 8. Plant Hormones. 10. β-indolyl acetic acid (IAA). 11. Gibberellic acids; cytokinins—kinetin, zeatin; ethylene; abscisic acid (dormin). 12. 2:4 Dichloro-phenoxyacetic acid (2:4 − D); phenylacetic acid; phenoxyacetic acid; α-naphthalene acetic acid; naphthoxyacetic acid. 13. Cell enlargement; cell division; bud dormancy; seed dormancy (germination); root initiation; onset of flowering; apical dominance in buds; cambial activity. 14, 15. Vol. II, Chap. 8. Auxins, Gibberellins and Cytokinins and Chap. 9. Translocation of Plant Growth Substances. 16. As root initiators in cuttings; preventers of fruit drop; selective weed killers. 17. A few parts per million.

Animals

1. By the transmission of electrical impulses; by the dispersion of soluble substances through the body fluids; by mechanical effects between parts. That is, nervous, hormonal and proprioceptive (kinaesthetic) co-ordination. 2. Vol. II, Chap. 8. The Endocrine Glands and Animal Hormones. 3. They regulate differentiation and growth; maintain the body in a steady state; maintain a balance of metabolism; are concerned with reproduction and care of young. 4. Organizers direct and control orderly general development in embryos by inducing some parts to develop. They are sometimes called evocators when localized and specific in action. A general organizer is produced by the dorsal lip of the blastopore in amphibia. Examples of more specific evocators are the substance produced by cells underlying the ectoderm on the dorsal side of a vertebrate embryo inducing or evoking those above to become neural plate and a substance that evokes eye lens development in the right place in the epidermis of vertebrates. 5. They are substances secreted into the external environment, acting like hormones to influence the behaviour of other animals,

especially those of the same species, with regard to social identification, integration of groups and sexual behaviour. In their effects possibly but they are not easy so to define because they are externally liberated and not effective in quite the same ways. 6. Vol. II, Chap. 8. Pheromones. 7. Brain and spinal cord; cranial and spinal nerves and autonomic nervous system; an organ designed to detect a stimulus of some kind, e.g. eye, ear. 8, 9, 10. Vol. II, Chap. 8. The Synaptic Nervous System. 11. The control of an activity that is automatic, not under conscious control, e.g. heart beat. 12. A response to stimulation that is evoked through an "inborn" nervous pathway, i.e. one independent of experience, e.g. knee jerk. See also Vol. II, Chap. 8. The Reflex Arc and its Complications. 13. Vol. I, Chap. 29. The Brain. Vol. II, Chap. 8. The Reflex Arc and its Complications. 14, 15. Vol. II, Chap. 8. Conduction of Impulses Through the Nervous System. 16. If a nerve impulse is strong enough to be propagated, the speed of conduction is independent of the strength of the stimulus. 17. (*a*) 2000 per second, under experimental conditions in man; (*b*) 160 metres per second in man. 18. Vol. II, Chap. 8. Conduction of Impulses Through the Nervous System. 19. It describes the effect of a nerve impulse (action potential) in increasing excitability of a nerve in advance of its own transmission. 20, 21, 22. Vol. II, Chap. 8. Transmission across Synapses and Junctions.

Translocation of Materials (p. 62)

1. Every cell is assured of a supply of respiratory substrate and oxygen; nutrients are transported to metabolic centres; every cell can eliminate wastes; growing zones and reproductive cells can be adequately supplied with new material; storage substances can be passed to storage tissues; secretions can be satisfactorily dispersed.

Plants

2. Nutrients; elaborated food materials; hormones. 3. Examination of exudates; ringing experiments; use of radioactive tracer substances. 4. Vol. II, Chap. 9. Transport of Nutrients. 5. Ringing experiments; extraction of phloem sap with aphid stylets. 6. They tend to move multidirectionally whereas minerals tend to move from the root upwards. 7. Phloem. 8. Vol. II, Chap. 9. Transport of Elaborated Food Material. 9. Mass flow due to pressure (water potential) differences at different levels, these caused by evaporation above and/or root pressure from below. 10. Vol. II, Chap. 9. Mechanism of Movement of Substances through the Xylem and Phloem. 11. Mass flow due to turgor gradient; diffusion; pressure-flow mechanism; cytoplasmic streaming. 12, 13. Vol. II, Chap. 9. Mechanism of Movement of Substances through the Xylem and Phloem. 14. Vol. II, Chap. 9. Translocation of Plant Growth Substances. 15, 16. Passively by diffusion. 17. Terrestrial plants could be successful only if they had adequate means of transporting materials within their own bodies from sites of intake or formation to other localities. A vascular system does this. Every cell of an aquatic can be in

continuous contact with the outside medium of the environment and hence be self sufficient without such an internal communication system.

Animals

1. By a diffusion process. 2. Porifera; Coelenterata. 3. It is carried by wandering amoeboid cells. 4. Vol. II, Chap. 9. The Use of External Water as a Translocating Agent. 5. They are dispersed through fluid filled cavities in the body by muscular activity and thence by diffusion into the adjacent cells. 6. Vol. II, Chap. 9. Translocation in Acoelomates. 7. By haemocoelic translocation in conjunction with blood vessel and a contractile heart forming an open vascular system. Vol. I, Chaps 22 and 23. The Blood Vascular System. Vol. II, Chap. 9. Haemocoelic Translocation. 8. Insects have heart pumping mechanisms correlated with tracheate external gaseous exchange systems. Cephalopods have an extra pair of hearts, hence more efficient pumping. 9. Respiratory exchange occurs via the water vascular system; food is distributed from the gut diverticulae by diffusion probably. Vol. II, Chap. 9. Translocation in Echinoderms. 10. Food materials; gases; excretory products; hormones. 11. By a heart pumping mechanism. 12. Neurogenic—the heart beat is initiated by incoming nerve impulses; myogenic—the heart beat is initiated by the heart itself. 13. Vol. II, Chap. 9. The Heart. 14. These nodes are groups of Purkinje tissue, extensions of which carry excitation. impulses to other parts of the heart. 15. The beat is initiated in the sinu-auricular node but the rate of beat is controlled by the nervous system. Vol. II, Chap. 9. The Heart. 16. There are differences in wall structure and in the manner and direction of conveying blood. Vol. II, Chap. 9. Blood Vessels. 17 Vol. II, Chap. 9. The Lymphatic Circulation. 18. The return of lymphatic body fluids to the heart; the replenishment of some blood cells (lymphocytes); the destruction of microbial invaders. 19. Vol. II, Chap. 11. Respiratory Pigments. 20. Haemoglobin—vertebrates, most annelids, some molluscs; haemocyanin—most molluscs, many crustaceans, some arachnids; chlorocruorin—polychaete annelids; haemoerythrin—sipunculids.

Nutrition (p. 63)

1. To supply the material from which energy can be made available and with which growth and repair can be achieved; for secretion processes; for storage requirements; for maintainance of internal osmotic and pH conditions. 2. Vol. II, Chap. 10. Introduction. 3. No. All protoplasmic systems require the same basic materials to serve equivalent purposes.

Plants

4. Carbon, hydrogen, oxygen, nitrogen, phosphorus, sulphur, potassium, magnesium, calcium and iron. 5. Boron, manganese, zinc, copper, molybdenum. These are required only in concentrations of a few parts per million, i.e. traces. 6, 7, 8. Vol. II, Chap. 10. Material Requirements: Their Sources

and Roles. 9. Artificial culture solution technique. Vol. II, Chap. 10. The Effects of Nutrient Deficiency—Water-culture Technique. 10. There is endless circulation of materials through protoplasmic systems and the environment. Living things extract from their surroundings but eventually die and decay releasing materials for further use in succeeding generations. 11. Vol. II, Chap. 10. The Nitrogen Cycle. 12. Nitrogen fixation; absorption of nitrogenous material from the soil and its eventual return; reduction of inorganic nitrate (denitrification). See also Vol. II, Chap. 10. The Nitrogen Cycle. 13. Vol. II, Chap. 10. The Nitrogen Cycle. 14. By manufacturing artificial nitrogenous fertilizer; by the mineralization of organic material from urban communities and the dispersal to sea of the products of his sewage installations. 15. Vol. II, Chap. 10. Carbon, Sulphur and Water Cycles. 16. The assimilation of carbon into the living system by the synthesis of carbon compounds from carbon dioxide and water. 17. Carbon assimilation using the energy of light. 18. Carbon dioxide availability; water availability; mineral salt availability; light conditions; temperature conditions. 19. Chlorophyll presence; the activity of the relevant enzymes. 20. Vol. II, Chap. 10. Carbon Assimilation by Photosynthesis and Factors affecting it. 21. Carbon dioxide intake by leaves measured as the volume of the gas absorbed per hour per unit area of leaf surface (allowance being made for carbon dioxide released by respiratory activity of the tissues); oxygen output by photosynthesising tissues of aquatic plants measured as the number of uniformly sized bubbles of gas emitted per unit time (useful only for comparing rates under different external conditions); organic material increase by tissues measured as an increase in dry weight per unit time per unit mass of photosynthesising tissue or grams of organic material elaborated per hour per unit area of leaf surface. 22. Vol. II, Chap. 10. Light and Chlorophyll in the Mechanism of Photosynthesis. 23. The construction of chemical bonds—ADP + inorganic phosphate $\rightleftharpoons$ ATP and NADP + 2H $\rightleftharpoons$ NADP.H_2. 24. Briefly, light is the energy source whilst chlorophyll acts as the energy collector or harnesser as the first step in making the light energy perform the work of transporting electrons "uphill" to effect the formation of ATP and NADP.H_2. See also Vol. II, Chap. 10. Z-scheme diagram. 25. Red and blue-violet. 26. Vol. II, Chap. 10. Light and Chlorophyll in the Mechanism of Photosynthesis. Chap. 11. Respiration Using a Carbohydrate Substrate. 27. Vol. II, Chap. 10. The Chemical Pathway. 28. $6\,CO_2 + 12\,H_2O \rightleftharpoons C_6H_{12}O_6 + 6\,H_2O + 6\,O_2$. 29. Water; in the light reaction. 30. Carbon dioxide is attached to an acceptor molecule, ribulose diphosphate. 31. Phosphoglyceric acid. 32. Vol. II, Chap. 10. The Chemical Pathway. 33. Amino acids and proteins; fatty acids and glycerine to give lipides. 34, 35, 36, 37. Vol. II, Chap. 10. The Synthesis of Other Compounds. 38. The bacterium uses H_2S as the source of reducing hydrogen instead of H_2O. 39. Fungi cannot photosynthesize and so live as heterotrophes on previously elaborated organic materials in the same way as animals do. 40. Chemosynthesis is the use of the bond energy of inorganic compounds as the energy source for carbon assimilation, cf. photo (light)—synthesis; the nitrifying bacterium, *Nitrosomonas* sp., performs the reaction—$2\,HNO_2 + O_2 \rightleftharpoons 2\,HNO_3 +$ Energy. 41, 42. Vol. II, Chap. 10. 43. Paper partition chromatography; use of radioactive tracers.

Animals

1. Carbohydrates; nitrogen-containing compounds (amino acids or proteins); fats; vitamins; mineral salts; water. 2. Vol. II, Chap. 4. 3, 4, 5, 6. Vol. II, Chap. 10. Material Requirements, Their Sources and Roles. 7. Some apparently, e.g. *Chilomonas* sp., can live in darkness with carbon dioxide as the sole source of carbon. 8. Ingestion—food intake to the alimentary canal; digestion—food rendered soluble by enzyme action (generally hydrolysis); absorption—food intake by gut cells; assimilation—food material incorporated into tissue substances; egestion—elimination of the unabsorbed portion of the food intake. See also Vol. II, Chap. 10. Treatment of the Food Material. 9. Pseudopodial by amoeboid protozoa; ciliary by ciliate protozoa; tentacular by some holothurians; mucoid by some gasteropods; setose by some aquatic crustaceans. 10. Taking in and sifting quantities of static material such as mud detritus by burrowing animals such as some annelids and some echinoderms; scraping and boring away parts of large masses as by some gasteropods, termites; seizing and chewing prey as by animals with toothed jaws and other special mouth parts. 11. Piercing and sucking animals such as leeches, ticks and fleas. See also Vol. II, Chap. 10. Feeding Mechanisms. 12. Vol. I. Animal Type Chapters. 13. Vol. II, Chap. 10. Digestion. 14. Extra-cellular digestion occurs in the gut cavity into which enzymes are secreted by adjacent cells and accessory glands; intra-cellular digestion occurs inside cells after food materials have been taken in. 15, 16. Vol. II, Chap. 10. Digestion. 17. Vol. II, Chap. 10. Absorption. Assimilation and Fate of Absorbed Substances. 18. The typhlosole of the earthworm; digestive glands of crayfish; gut caeca of insects; spiral valve of dogfish; villi on the gut wall of mammals. 19. Undigested food residues in the gut with other matter that may be discharged into it. See also Vol. I. Animal Type Chapters.

Respiration (p. 66)

1. They are equivalent; both expend energy in the performance of work. 2. As energy of chemical bonds; in the long term in respiratory substrates (most commonly fats and carbohydrates), in the short term as bond energy in ATP. 3. To maintain the protoplasmic system as an integrated whole; to do chemical, mechanical and osmotic work; to energize electrical impulse transmission and to maintain electric potential differences across membranes; sometimes also heat release and sound and light emission. 4. The overall process by which energy is made directly available for work inside cells. This energy, immediately available, is in the form of chemical bonds in ATP. 5. Breathing relates to the external exchange of gases that may be essential to the internal energy release processes. 6. A substance from which chemical bond energy can be made to do the work of combining inorganic phosphate and ADP to produce ATP, the immediate energy source for the cell. 7. Glucose—2.8×10^3 kJ mol^{-1}; fat (average)—35.5×10^3 kJ mol^{-1}. 8. The gaining of positive charge by an element or compound by the loss of electrons. See also Vol. II, Chap. 11 and Appendix. Oxidation and Reduction. 9. By the addition of oxygen; the removal of hydrogen; the removal of electrons. 10. The loss

of positive charge by an element or compound by gain of electrons—the reverse of oxidation. 11. $C_6H_{12}O_6 + 6\,O_2 \rightleftharpoons 6\,CO_2 + 6\,H_2O + 2.8 \times 10^3\,kJ$. 12. Vol. II, Chap. 11. Biological Oxidation and Reduction. 13. The oxidation-reduction enzymes—dehydrogenases, oxidases, peroxidases and catalase. 14. Vol. II, Chap. 5. Oxidation and Reduction Enzymes. ' 15. Dehydrogenases. 16. Substances to which hydrogen can be transferred in biological oxidations. 17. Nicotimamide adenine dinucleotide (NAD or Coenzyme 1); flavin adenine dinucleotide (FAD); flavin mononucleotide (FMN); the cytochromes.

$$\underset{\substack{\text{reduced}\\\text{substate}}}{AH_2} + \underset{\text{hydrogen acceptor}}{NAD} \xrightarrow{\text{dehydrogenase}} \underset{\substack{\text{oxidized}\\\text{substrate}}}{A} + \underset{\substack{\text{reduced hydrogen}\\\text{acceptor}}}{NAD.H_2}$$

18. $NAD.H_2 + FAD \xrightarrow{\text{dehydrogenase}} NAD + FAD.H_2;$

$FAD.H_2 + \text{cytochrome (oxidized)} \xrightarrow{\text{dehydrogenase}} FAD + \text{cytochrome}.H_2;$

$\text{cytochrome}.H_2 + O \xrightarrow{\text{cytochrome oxidase (cyt } a_3)} \text{cytochrome (oxidized)} + H_2O.$

19. It is used as the final hydrogen acceptor under aerobic conditions. 20. Vol. II, Chap. 11. Methods of release of Energy: Aerobic and Anaerobic Conditions. 21. Vol. II, Chap. 11. Respiration Using A Carbohydrate Substrate. 22. The aerobic stages (Krebs' cycle) follow the glycolytic breakdown if oxygen is available. 23. Glycolysis produces pyruvic acid with eight ATP bonds per glucose molecule respired; Krebs' cycle produces carbon dioxide and water with a further thirty ATP bonds per molecule of pyruvic acid respired. 24. Aerobic respiration is much more efficient than anaerobic. See also Vol. II, Chap. 11. Details of Respiration with a Carbohydrate Substrate. 25. Heat. 26. By completing the oxidation processes as a series of finely graded steps, avoiding great local changes in temperature. 27. Photosynthetic processes "capture" energy from light and store it as chemical bond energy in organic compounds—respiratory processes make this store energy available for work in cells. 28. It is equivalent to the glycolytic processes under anaerobic conditions but the pyruvic acid is further degraded to ethyl alcohol and carbon dioxide. 29. Oxidative phosphorylation is the formation of ATP from ADP and inorganic phosphate using the chemical bond energy of respirable compounds; photophosphorylation is the equivalent process using light energy. 30. Respiratory quotient (R.Q.) is the ratio of carbon dioxide evolved to oxygen absorbed during respiration of any given substrate. See also Vol. II, Chap. 11. Respiratory Quotients. 31. The possible nature of tissue respiration. 32. Vol. II, Chap. 11. Gaseous Exchanges at Respiratory Surfaces. 33, 34, 35. Vol. I. Animal Type Chapters. 36. Highly vascular epidermal tissue as in annelids; gills, haemocoelic as in crustaceans and vascularized as in fishes and tadpole stages in amphibians; trachaea as in insects; pulmonary chambers as in molluscs; lung books as in spiders; lungs as in birds and mammals. 37. Vol. II, Chap. 11. Respiratory Pigments. 38. They are not photosynthesizing

tissues and thus no account need be taken of this; they are very easy to handle under practical conditions.

Growth and Development (p. 68)

1. The growth of a population is a change in numbers of individuals whilst the growth of an organism is a change in its bulk of living substance (this latter is difficult to determine precisely). 2. Vol. II, Chap. 13. Growth of a Population. 3. $N_{t_2} = N_{t_1} . e^{kT}$. See also Vol. II, Chap. 13. Growth of a Population. 4. The exponential law. 5. They are basically the same. 6. Vol. II, Chap. 13. Growth of a Population. 7. A sigmoid curve. It ceases to obey the exponential law. 8. k is decreasing, that is, the efficiency with which new growth is made continuously falls away. 9. Material requirements are being used up; metabolic by-products, possibly toxic, interfere with normal metabolism. 10. In principle, their patterns of population growth are the same. 11. In terms of changes in its active protoplasmic substance. 12. It is inextricably bound to inert (non-living) material in the body. 13. A linear dimension; dry weight. 14. Change in size need not reflect true growth; dry weight changes in the same individual cannot be studied.

Plants

15, 16, 17, 18. Vol. II, Chap. 13. Growth of Multicellular Organisms. 19. The form of the growth pattern; the rate of growth; the relative growth rate. 20. Relative growth rate is given by the ratio between the rate of growth at any time and the amount of growth already made. It measures the efficiency with which an organism can add more substance per unit substance already there. 21–25. Vol. II, Chap. 13. The Interpretation of Growth Measurements. 26. Vol. I. Plant Type Chapters. 27. Nutrient availability; accumulation of by-products of metabolism; temperature; light conditions; pH conditions. 28. Vol. II, Chap. 13. Factors Affecting Growth. 29. The effect of keeping a green plant continuously in darkness. 30. Hormonal conditions. 31. Vol. II, Chap. 13. Regeneration, Wound Healing and Tissue Replacement in Plants. 32. Vol. II, Chap. 8 and Chap. 13 as for 31. 33, 34. As for 31. 35. Bark formation. 36. Vol. I, Chap. 17. Secondary Growth of the Stem. 37. Inherited internal influences (genes); conditions of the external environment. 38. Zygote to embryo; dormancy period (if any) in the seed; seed germination; vegetative development; reproduction. 39. Water and oxygen must be available; the temperature should be within the proper limits; light conditions fulfilled sometimes (see photoblasty). The embryo must be fully formed and "after-ripening" completed if necessary. 40. Imbibitional and osmotic forces. 41. To convert protoplasm to its active hydrated state; for the hydrolysis of food reserves; to act as solvent for metabolites. 42. It serves an essential respiratory function, i.e. it is the final hydrogen acceptor in the oxidation processes required to make energy available within cells. 43. Carbon dioxide is evolved; the temperature of the tissue rises. 44. Low and high temperatures inhibit through effects on enzyme activity. 45. Light indifferent—germination not affected by light conditions, i.e. not photoblastic, as in most plants; light sensitive—will not germinate unless exposed to light for at least a short period, i.e. positively photoblastic, as in species of *Veronica* and *Lythrum*; light hard or negatively photoblastic—germination is retarded by exposure to light, as in species of *Phlox* and *Allium*. 46. Vol. II, Chap. 13. Germination and Dormancy. 47. Morphological immaturity of the seed; testa impermeable to water and/or oxygen; the seeds have an "after-ripening" requirement. 48. Possibly some chemical and/or physical change occurring in the seed under the influence of internal factors. 49. Changes in permeability of the seed coat. 50, 51. Vol. II, Chap. 13. Germination and Dormancy. 52. Vol. I, Chap. 17. General Vegetative Organization. 53. It commences with the formation of flower primordia at an apex and ends with fruit ripening. 54. The state of vegetative maturity of the plant; the relative lengths of alternating light and dark periods to which the plant has been subjected; fulfilment of a possible chilling requirement. 55. Response to alternating light and dark periods. See also Vol. II, Chap. 13. Phase of Reproduction. Chap. 17. Sexual Reproduction in Plants. 56. Treatment at low temperature or chilling. See also Vol. II, Chap. 13. Phase of Reproduction. 57. On pollination being successfully completed followed by fertilization of an ovum in an ovule. 58. Vol. II, Chap. 13. Phase of Reproduction. 59. Vol. II, Chap. 8. Plant Hormones. Chap. 13. Phase of Reproduction. Chap. 17. Sexual Reproduction in Plants —Onset of Flowering. 60. Vol. II, Chap. 8. Plant Hormones.

Animals

1–5. Vol. II, Chap. 13. Growth of Multicellular Organisms. 6. Skin; gonads; bone marrow. 7. External—nutrient availability; accumulation of metabolic by-products; temperature; light conditions; pH conditions: internal hormones. 8. The process by which an organism regains its normal form when this has been altered by loss of a part. 9, 10. Vol. II, Chap. 13. Regeneration, Wound Healing and Tissue Replacement in Animals. 11. Zygote to embryo (gestation); release of embryo; attainment of mature form and sexual maturity; reproductive phase; senescence. 12. Cleavage and formation of blastula; gastrulation; organogenesis. See also Chaps. 27, 28 and 29 of Vol. I. Development. 13. Vol. II, Chap. 13. Development of the Vertebrate Animal. 14. Vol. II, Chap. 13. Hatching: Birth. 15. Vol. II, Chap. 13. Attainment of Sexual Maturity. 16. By hormone control. 17. Senescence is the ageing effect produced by degenerative processes such as cells losing their capacity to divide and changes in hormal balance. 18. The study of the causes of senescence or aging. It may be important in prolonging the active, vigorous period of life.

Locomotion (p. 71)

Animals

1. Amoeboid; ciliary or flagellar; gregarine (myoneme). 2. Vol. II, Chap. 12. Locomotion in Animals. Vol. I, Chaps. 6, 7, 8, 32. 3. Vol. II, Chap. 12. Locomotion in Animals. 4. Individual cells of higher animals may make

amoeboid or flagellar (ciliary) movements. 5. *Euglena*; *Monocystis*; *Stentor*; *Vorticella*. 6. Vol. I, Chap. 10. *Hydra*. Vol. II, Chap. 12. Musculo-epithelial cells. 7. Skeletal and cardiac (both striated) and smooth (unstriated). 8. Vol. I, Chap. 20. Muscular Tissues. 9. Vol. II, Chap. 12. Muscular movement. 10. Vol. II, Chap. 8. Transmission Across Synaptic Junctions. 11, 12, 13, 14. Vol. II, Chap. 12. Muscular Movement. 15. An accumulation of lactic acid occurs when the oxygen supply to the tissue is inadequate for the complete oxidation of glucose. 16. Vol. II, Chap. 12. Modes of Progression by Muscular Contraction. 17. Flying fish; flying frog; flying lizard; flying snake; gliding opossum; flying squirrel. 18. Vol. II, Chap. 12. Ciliary and Flagellar Movement. Muscular Movement. 19. Vol. I, Chap. 20. Connective Tissues. Vol. II, Chap. 12. Muscular Movement. 20. Muscles of the iris; ciliary muscle of the eye lens. 21. Cheetah about 110 km per hour; sailfish, 105 km per hour; swift, 350 km per hour; dragonfly, 80 km per hour.

Plants

1. Bacteria; motile green algae such as *Chlamydomonas* sp.; diatoms such as *Pinnularia* sp.; slime fungi such as *Badhamia* sp. 2. Bacteria and algae by flagellar movement; diatoms by streaming cytoplasm mechanism; slime fungi by a form of amoeboid movement. 3. The male reproductive cells or antherozoids of bryophytes (*Funaria*), pteridophytes (*Dryopteris*) and cycads (*Cycas*). 4. The aquatic environment; the presence of liquid water.

Secretion and Storage (p. 72)

1. A cell secretes when it separates (internally or externally) materials from the actively metabolizing protoplasm so that they become independently recognizable. Excretion is defined as the elimination of unwanted, often toxic, by-products of metabolism by a secretory process. 2. Intra-cellular—secretory substance separated internally to the cell, e.g. fat as internal glubules; extra-cellular—secreted substance separated externally to the cell, e.g. digestive juices passed to the gut cavity. 3. A storage substance, e.g. starch, is later metabolized and made use of. A permanent secretion remains unchanged indefinitely, e.g. cellulose walls of plant cells, not normally being capable of transformation. 4. For the survival of adverse conditions; the endowment of a new generation with the requirements for initial growth and development.

Plants

5. Starch grains; sugary substances; resin; cutin; latex. 6. Storage parenchyma in root or stem—stored food; nectary surface—insect attraction; resin ducts—possibly a protective function; epidermal cells—water-proofing; laticiferous vessels—storage. 7. Vol. II, Chap. 14. Secretory Processes. 8. To provide an initial material and energy supply for renewed activity after a period of "rest"; to provide the requirements for the initial growth activity of an embryo. 9. Vol. I, Chap. 17. Seed Structure. 10. Vol. I, Chap. 17. Modifications of Roots; of Stems; of Leaves. 11. Starch; glycogen; inulin; sucrose;

glucose; hemicellulose. 12. Vol. II, Chap. 14. Storage Substances. 13. Fats are stored as oil globules most commonly in reproductive structures such as spores and in the endosperm tissue of seeds, e.g. castor oil. 14. Protein granules such as aleurone grains are commonly found in seeds; amides such as asparagine and glutamine are sometimes found in storage parenchyma.

Animals

1, 2. Vol. II, Chap. 14. Functions and Nature of Secretory Substances. 3. Buccal cavity from salivary glands; stomach from gastric glands; duodenum from Brunner's glands and crypts of Leiberkuhn and via the pancreatic duct from the pancreas—but all cells must manufacture enzymes internally in order to catalyse metabolic reactions within themselves. 4. Vol. II, Chap. 10. Digestion. 5. Vol. II, Chap. 14. Secretory Processes by Animal Cells. 6. Exocrine secretions are released into cavities; endocrine secretions are released into body fluids (blood stream); holocrine secretions are produced by cells or glands, the substance of which forms part of the secretion. 7. Nervous control —salivary glands, initial gastric juice secretion by the stomach, some bile secretion by the liver; hormonal control—sustained gastric secretion, duodenal secretion, pancreatic secretion. 8. Corals; molluscs; crustaceans. 9. Silica by sponges; strontium sulphate by radiolarians; enamel by mammalian teeth and fish dermal denticles; silk by spiders; wax by honey-bees. 10. To tide an animal over adverse nutritional conditions; to form reserves for the initiation of activity in young animals, e.g. yolk. 11. Glycogen; rarely others on a long term basis. 12. Muscle and liver. 13. Insects—fat body; fish—liver; amphibian—liver and fat bodies; reptiles—fat bodies; mammals—mesenteries and under the skin (adipose tissue). 14. Fat has a greater energy yield per unit mass and is therefore conservative of space.

Excretion (p. 73)

Animals

1. Vol. I. Animal Type Chapters. 2. Ammonotelic—ammonia is diffused directly into the surrounding medium as in many smaller invertebrates; uricotelic—ammonia is converted into uric acid and eliminated as in insects, birds, reptiles; ureotelic—ammonia is converted to urea and eliminated as in mammals. 3. In the liver. 4. Vol. II, Chap. 15. Nutrients in Excess of Requirements—The Ornithine Cycle. 5, 6. Vol. I. Animal Type Chapters. 7. Uric acid in the fat bodies of insects; calcium carbonate deposits in earthworms; crystalline bodies in plants. 8. Vol. I. Animal Type Chapters. 9. Vol. II, Chap. 15. Excretory Methods in Animals. Vol. I. Animal Type Chapters. 10. Vol. II, Chap. 15. Mammalian Kidney Function. 11. Ultrafiltration is filtration under hydrostatic pressure; resorbtion is the extraction of fluid back into the cells of the kidney tubule; osmoregulation is the adjustment of the cell and body fluid concentration by regulation of the water content.

Plants

1. Vol. II, Chap. 15. Excretory Methods in Plants.

Sensitivity and Response (p. 74)

1. Sensitivity is the capacity to perceive and respond to changes in conditions of the environment; a stimulus is a change in conditions pronounced enough to elicit a response; a response is any change of activity; behaviour is the sum of all activities. 2. Vol. II, Chap. 16. Stimuli. 3. Through parts that function as perceptive structures. 4. Animals have usually clearly defined organs of perception, plants have not (except some very primitive forms such as free-swimming algae). Animal responses are of very wide variety, affecting most aspects of behaviour; plant responses are mostly growth and development responses, again excepting the free-swimming forms, these tending to respond in the manner of animals.

Plants

5. Changes in—movement of free-swimmers; growth direction of parts of fixed plants; developmental pattern; metabolic activities. 6. Paratonic—responses made to external stimulation, e.g. a growth curvature due to a change in direction of the incident light; autonomic—responses made to internal changes in conditions, e.g. cytoplasm may start to stream in a cell apparently quite unconnected with external stimulation of any known kind. 7. Mechanical movements; movements in response to stimulation of irritable protoplasm. 8. Vol. II, Chap. 16. Plant Movements in Response to External Stimuli. 9. Tactic movements are made by a whole organism or a freely locomotive part, the direction of the stimulus having a direct bearing on the direction of the response; tropic movements are made by parts of fixed plants, the direction of movement being governed by the direction of the stimulus; nastic movements are made by parts of fixed plants in response to diffuse, non-directional stimulation. 10. Phototaxis by free-swimming algae; chemotaxis by bacteria and antherozoids; rheotaxis by plasmodia of slime fungi. 11. Plants move towards or away from particular illumination conditions; towards or away from the source of a stimulating substance; movement is related to the direction of flow of a stream of water. 12. Geotropism by roots and shoots of vascular plants, parts of fungi (particularly sporophores), parts of bryophytes, algal holdfasts; phototropism by shoots and leaves of vascular plants, parts of fungi (particularly sporophores), parts of bryophytes, algal holdfasts; chemotropism by pollen tubes, fungal hyphae; haptotropism by tendrils of vascular plants; hydrotropism by plant roots, some fungal hyphae. 13, 14. Vol. II, Chap. 16. Plant Movements in Response to External Stimuli. 15. By use of the clinostat; by use of a spinning disc to substitute a centrifugal force for that of the earth's gravitational pull. 16. Decapitation treatment; use of Czapek's glass covers. 17. By a root zoning technique. 18. A growth curvature, due to differential growth rates on opposite sides of the responding organ. 19. Vol. II, Chap. 16. Geotropism. 20. Vol. II, Chap. 16. Phototropism. 21.

By using a unidirectional light stimulation technique; coleoptiles. 22. By a capping or decapitation treatment. 23. By zoning the tip and locating the region of bend after a response has been made. 24. A growth curvature, due to differential growth rates on opposite side of the responding organ. 25, 26. Vol. II, Chap. 16. Phototropism. 27. Vol. II, Chap. 16. Other Tropic Responses. 28. Nyctinasty made as a "sleep" movement of some leaves related to change from day to night (may be a photonasty) e.g. leaves of *Oxalis acetosella*; thermonasty made as the opening or closing of some flowers related to changes in temperature, e.g. crocus flowers; seismonasty made as changes due to some physical shock treatment, e.g. the leaf movements of the sensitive plant, *Mimosa pudica*; haptonasty made as a response to contact, e.g. movements of stamens of *Centaurea* sp. and of the leaves of the insectivorous *Dionaea muscipula*; chemonasty made as a response to the presence of chemicals, e.g. the "tentacles" of the sundew leaf. 29, 30, 31. Vol. II, Chap. 16. Nastic Movements. 32. A nutational movement (circumnutation). 33. Vol. II, Chap. 16. Autonomic Movements. 34. Vol. II, Chap. 16.

Animals

1. Movements of whole animals (locomotion); movements of parts; secretory activity; colour changes; changes in metabolism. 2. An exteroceptor receives stimulation from sources external to the animal; an enteroceptor is stimulated by changes in conditions internally (the meaning may be restricted to changes in gut and respiratory cavities); a proprioceptor receives stimulation from changes within tissues, detecting alterations in tensions occasioned by changes in position and movements within them. 3, 4. Vol. I. Animal Type Chapters. 5. Vol. II, Chap. 16. Gustatory Sense. 6. Vol. I. Animal Type Chapters. 7. Vol. II, Chap. 16. Photoreception. 8. Deuteranopia is an inability to distinguish properly green from red (green blindness); protanopia, less sensitive still to reds than in deuteranopia (red blindness); tritanopia, an inability to distinguish between yellow and blue (blue-violet blindness). 9. Vol. II, Chap. 16. CNS Visual Mechanism. 10. Diurnal animals have more cones than rods in the retina; possess a fovea; have greater lens-retina distance; possess colour filtering substances. Nocturnal animals have more rods than cones; have slit pupils; have a reflecting layer in the choroid. 11. Protection from drying; difference in shape of the cornea. 12. Vol. II, Chap. 16. Vertebrate Visual Adaptations. 13. Frequency; amplitude (intensity of sound or loudness); direction of source. 14. The first by the organ of Corti in the cochlea; the second by the quantity of nerve impulses reaching the brain; the last by a phase difference between the sound waves from the same source reaching the separate ears helped possibly by differences in their intensities. 15. Balance and general body equilibrium. 16. Organs of the lateral line; found in fish and some amphibian larvae. 17. They detect vibrations transmitted through water. 18. Vol. II, Chap. 16. Lateral Line System of Fish. 19. By the swim bladder. See also Vol. II, Chap. 16. Effect of Hydrostatic Pressure on Fish. 20. Vol. II, Chap. 16. Effect of Atmospheric Pressure on Land Vertebrates. 21. Touch; temperature (heat) changes; pain. 22. Tactile corpuscles; end bulbs; Pacinian corpuscles; organs of Ruffini. 23. As a

warning device. 24. Vol. I. Animal Type Chapters. Vol. II, Chap. 16.
Equilibrium Perception. 25. Pigment cells; leeches; stick insects; octopus;
prawn; frog; chamaeleon; human being. 26. Vol. II, Chap. 16. Colour
Change Responses. 27. Light; temperature; touch; humidity. 28. Vol. II,
Chap. 16. Colour Change Responses. 29–36. Vol. II, Chap. 16. Animal
Behaviour.

Reproduction (p. 77)

1, 2, 3. Vol. II, Chap. 17.

Plants

4. Attainment of sexual maturity or onset of flowering; gametogenesis;
liberation of gametes; fertilization and zygote formation. 5. By the development of flowers. 6. Its state of vegetative development (internal) and temperature light conditions (external). 7, 8. Vol. II, Chap. 8. Plant Hormones.
Chap. 17. Onset of Flowering. 9. Vol. II, Chap. 17. Attainment of Sexual
Maturity or Onset of Flowering. 10. Tuber formation; branching; dormancy.
11, 12. Vol. II, Chap. 17. Onset of Flowering. 13. (*a*) Vol. I, Chap. 16. Development of Pollen Grains. Development of the Ovule. (*b*) Vol. I. Plant
Type Chapters. 14. Meiotic (reduction division). 15. Vol. I. Plant Life Cycle
Diagrams in Type Chapters. 16. Vol. I, Chap. 2. Meiotic Division of Cells.
17. Vol. I. Plant Type Chapters. 18. Vol. I, Chap. 16. Fertilization. See also
Vol. II, Chap. 17. Fertilization and the Zygote. 19. Vol. I. Plant Type
Chapters. 20. Zoidogamy occurs when gametes are brought together by the
free-swimming movements of at least one of them; siphonogamy occurs when
gametes are brought together by growth of a tube through which one of them
is conveyed to the other. 21. Zoidogamy to aquatic and siphonogamy to
terrestrial (dry) conditions. 22. Chemical stimulation; they make chemotactic
responses. 23. Vol. I, Chap. 16. Fertilization. 24. The zygotic nucleus will be
equivalent to a parent nucleus, i.e. diploid. 25. Through the genetic code
(DNA) carried in the chromosomes. 26. Because the genetic code it carries
is not very likely to be precisely the same as that of either parent. 27. When
both parents were genetically identical and the zygote was therefore genetically
identical with them. 28. It creates "material" for natural selection to operate
upon hence the "survival of the fittest". 29. It provides the means of spreading
advantageous features through a population. 30. Some risk is attached to the
attainment of effective union of gametes, thus much material may be wasted.
31. Reproductive processes have evolved tending to follow patterns (according
to conditions of the environment) always towards effective fertilization with
the least wastage. 32. By vegetative propagation and by formation of specialized (but non-gametic) cells such as spores. 33. Vol. I. Plant Type Chapters.
34. It has no requirement for fusion with another equivalent structure before
further development. 35, 36. Vol. I. Plant Type Chapters. 37. No. 38.
Vegetative propagation. Grafting; taking of cuttings; budding; layering,
etc. 39. Sexual reproduction making use of selected parents to attempt to
create a specific genetic composition of a zygote.

Animals

1. Attainment of sexual maturity; gametogenesis; liberation of gametes; fertilization. 2. Gonadotropins. 3. Vol. I. Animal Type Chapters. See also Vol. II, Chap. 17. Attainment of Sexual Maturity. 4. The gonadial hormones (androgens and oestrogens). 5. Changes in hormonal balance. See also Vol. II, Chap. 17. Attainment of Sexual Maturity. 6. Female—primordial germ cell, oogonium and primary follicle, primary oocyte and Graafian follicle, first meiotic division, secondary oocyte with first polar body, second meiotic division, ovum with second polar body. Male—primordial germ cell, mitotic divisions, spermatogonia, primary spermatocytes, first meiotic division, secondary spermatocytes, second meiotic division, spermatids, spermatozoa. 7, 8. Vol. II, Chap. 17. Spermatogenesis. Oogenesis. 9. Vol. I. Animal Type Chapters. 10. By contraction of the gubernaculum. 11. Earthworm; cock. roach; snail. 12. To resist further penetrations. 13, 14. Vol. II, Chap. 17- Animal Fertilization. 15, 16. Vol. II, Chap. 17. Reproductive Rhythms· 17, 18. Vol. I. Animal Type Chapters. See also Vol. II, Chap. 17. Care of Young. 19. *Amoeba, Euglena*; *Sarcocystis, Plasmodium*. 20. Plasmotomy, in which a multinucleate mass of cytoplasm separates into smaller pieces without nuclear divisions; reported to occur in *Opalina*. 21. Separation of the body into unspecialized parts as in some coelenterates and some aquatic annelids. 22, 23, 24. Vol. II, Chap. 17. Asexual Reproduction in Animals. 25. Development of a female gamete (or a cell functioning as one) into a new organism without being fertilized, e.g. rotifers, aphids, water fleas (*Daphnia*). 26. Exposure to hypo- and/or hypertonic salt solutions; changes in pH of surroundings; radiation with infra-red and ultra-violet radiations; needle puncture. 27. Distribution of the sex (X and Y) chromosomes. See Vol. II, Chaps 17 and 19. Sex Inheritance and Sex Linkage. 28. Vol. II, Chap. 17. Intersexes and Sex Reversal. 29. An incomplete separation of the sexual characteristics in an individual. See also Vol. II, Chap. 17. Intersexes and Sex Reversal. 30. (*a*) The oyster alternating male and female; (*b*) all male frogs from eggs incubated at 27°C before fertilization, irrespective of chromosome make-up; (*c*) castrated male toads on fat rich diet become functional females. 31. Showing male characters in some parts of the body and female fn others as in some insects. 32. The sex chromosomes that determine the form of the gonads and the sex-determining substances (hormones) that these produce.

Special Modes of Life (p. 79)

1, 2. Vol. I, Chap. 4. Modes of Life. 3. Radiant energy as light and chemical bond energy. 4. Inorganic "raw" materials and elaborated organic substances. 5. Plants use light energy and inorganic materials; animals use the organic materials to supply both energy from chemical bonds and the food substances. 6. Photoautotrophes use light, e.g. chlorophyllous plants and bacteria; chemoautotrophes use chemical bond energy of inorganic substances, e.g. a few bacteria such as the nitrifying *Nitrosomonas* and *Nitrobacter*. 7. Fungi and many bacteria. They use already elaborated organic materials as sources of energy and nutrients. 8. Free-livers feed freely from among other living things, e.g. many animals; saprophytes use as food source dead tissue or organic product of another organism; parasites use as food source the living tissue or food substances of another living organism, it host, causing it some disadvantage. 9. Vol. I, Chap. 4. Modes of Life. 10. An association between living partners resulting in mutual advantage(s). 11. It is said to occur when one organism derives benefit from an associate without any apparent beneficial or injurious effects to it.

Parasitism (p. 80)

12. Vol. I, Chaps. 31 and 32. 13, 14, 15, 16. Vol. I, Chap. 30. Parasitism. Characteristics of Parasites. 17. Vol. I, Chaps. 31 and 32. 18. To eliminate or reduce their deleterious effects on himself or on his possessions. 19. To know all about the parasite in detail. 20. Elimination of susceptible hosts; destruction of the parasites on or in the host; destruction of the parasites during transmission between hosts. 21. Vol. I, Chap. 30. Control of Parasites. 22. Vol. I, Chaps 31 and 32. Examples—*Puccinia, Cuscuta, Plasmodium, Fasciola, Pulex.* 23. Vol. I, Chaps 31 and 32. Effect on Host. 24. Vol. I, Chap. 30. Tolerance of Host.

Saprophytism (p. 80)

25. Vol. I, Chap. 33. Saprophytic Plants. 26. Vol. I, Chap. 30. Saprophytism. 27. Putrefaction (decay) of substrates is a digestive process effected by saprophytes. 28. Vol. II, Chap. 10. Cycles of Raw Materials in Nature. 29, 30. Vol. I, Chap. 30. Saprophytism. 31. By producing substances toxic to themselves. 32. By destroying or degrading his stored materials and by being useful in some of his activities, e.g. brewing. 33. Vol. I, Chap. 33. 34. He can use yeast, for example, in brewing and bread-making. 35. Vol. I, Chap. 33. Examples—*Mucor, Aspergillus, Saccharomyces, Agaricus.*

Symbiosis (p. 81)

36. Vol. I, Chap. 34. Examples—alga-fungus in lichen; turbellarian—alga; termite—trichonymphid; legume—bacterium in root nodules. 37, 38. Vol. I, Chap. 30. Symbiosis. 39. Vol. 1, Chap. 34.

Commensalism (p. 81)

40. Vol. I, Chaps 30 and 35. Commensalism. Symbiosis implies mutual benefit; commensalism is now used to describe nutritional associations in which one partner may gain a positional advantage, say, without harming the other. 41. (*a*) Gut bacteria in animals; (*b*) epiphytic plants.

Insectivorous Plants (p. 81)

42. They couple autotrophism with heterotrophism, the latter by using the digestion products of small animals as food material. 43. They have greater access to nutrients that may otherwise be scarce in their habitats. 44. They are structurally adapted for catching prey and digesting the bodies externally to their own prior to absorbing the products. 45. Vol. I, Chap. 35. Examples— *Drosera, Utricularia, Pinguicula, Dionaea, Nepenthes, Sarracenia.*

Variation (p. 81)

1. Variation is the range of deviation from a standard or average condition. 2. It is an outstanding feature. 3. It ranges from the distinction between members representing large groups of living things, e.g. fungus and flowering plant; earthworm and mammal, to differences between any two individuals of the same species. 4. Investigations of the properties in which there are differences between organisms; classification in accordance with observable variations; a study of the causes of variation. 5. Compilation of an acceptable system of classification and identification. 6. Continuous variation exists in an ungraded manner through the range; discontinuous variation is that exhibited by a series of graded individuals, not merging into one another. 7. Environmental influences and parental (inherited) effects. 8. A mathematical study of measurable differences between organisms. 9, 10, 11, 12. Vol. I, Chap. 18. Measurement of Variation. 13. The mean value is the average value for all individuals; the mode value is that held by most individuals. 14–19. Vol. II, Chap. 18. Measurement of Variation. 20–22. Vol. II, Chap. 18. Inherited and Non-inherited Variation.

Genetics (p. 82)

1. Genetics seeks to account for the resemblances and differences exhibited among organisms that are directly related by descent. 2. The manner in which true-breeding variations within a species were related to one another. 3. Vol. II, Chap. 19. The Work of Gregor Mendel. 4. F1—all of one character (the dominant); F2—3 with dominant character to 1 with recessive character. 5. Vol. II, Chap. 19. The Work of Gregor Mendel. 6. Because body cells carry factors subscribed from both parents, i.e. two factors for each character. 7. Of a pair of contrasting characters only one can be represented in a single gamete. The law of the purity of the gametes. 8. When gametes are formed the pairs of factors representing characters separate from one another, a single gamete carrying only one of them. 9. If two coins are tossed simultaneously the frequencies of fall of two heads, head–tail or tail–head and two tails are 1:2:1. 10. F1—all showed a combination of the two dominant characters; F2—showed a segregation into a 9:3:3:1 ratio. 11. Vol. II, Chap. 19. The Work of Gregor Mendel. 12. Each of a pair of contrasted characters may be combined with either of another pair. The law of independent assortment. 13. For two or more pairs of characters the factors carried in a gamete can be recombined in all possible ways. 14. Each inheritable character is represented

in a gamete in some form; for each pair of contrasting characters, only one can be represented in the gamete and that for two or more pairs of such characters each representative in the gamete acts independently of the others. 15. He selected parents for his crosses so that they came from lines the ancestry of which was fully known. 16. Vol. II, Chap. 19. Terms Used in Genetics. 17. Eight. 18. Four; 9:3:3:1. 19. Incomplete dominance; multiple factors (polygenes); complementary factors; inhibiting factors; lethal characters. 20. Vol. II, Chap. 19. Apparent Deviations from Mendelian Inheritance. 21. Vol. II, Chap. 19. Material Basis for Mendelian Inheritance. 22. Many genes must be represented on each chromosome. 23. Linkage. 24. Certain genes will always tend to segregate together instead of acting independently of one another. 25–31. Vol. II, Chap. 19. Linkage and Crossing Over. 32. Vol. II, Chap. 19. Sex Inheritance. 33. Haemophilia; transmitted through mother to sons. 34. Vol. II, Chap. 19. The Material Basis for Mendelian Inheritance: The Chromosome Hypothesis. 35. The sudden appearance of recessive characters and changes in the genetical constitution of an individual. 36. The latter. 37. Mutations. 38. Through chromosome and gene mutations. 39, 40. Vol. II, Chap. 19. Chromosome Mutations. 41. Deficiencies; duplications; translocations; inversions. 42. White eye in *Drosophila*. Mutations may be induced by irradiation with X-rays, α-rays, β-rays, γ-rays, etc.; high temperature treatment; treatment with mutagenic substances. 43. Plasmagenes describes genetic material in the cytoplasm as separate from that in the nucleus. 44. Vol. II, Chap. 19. Hybrid Vigour. 45. Vol. II, Chap. 2. Chromosomes. Chap. 4. Purines, Pyrimidines and Their Derivatives. Chap. 19. Modern Concepts of the Gene and Gene Action.

Evolution (p. 86)

1. The study of fossils to gain knowledge of the origins of living things. 2. Archaeozoic, ended about 1100 million years ago; Proterozoic, about 500 million years ago; Palaeozoic, about 200 million years ago; Mesozoic, about 60 million years ago; Cenozoic, about 1 million years ago; Quaternary, to present. 3. Vol. II, Chap. 20. The Age of the Earth. 4. The preserved remains of a once living thing or an object that can be used as evidence of its existence. 5. Vol. II, Chap. 20. Fossils. 6. Preservation of undecayed remains such as insects in amber, mammoth in ice; impressions or prints such as leaves; moulds and casts such as *Calamite* pith casts, mollusc shell moulds and casts; petrifactions such as coal balls, sponges and corals; coprolites or faecal pellets; foot prints of animals. 7, 8, 9. Vol. II, Chap. 20. Living Things Through the Ages. 10. That living things of the present form parts of a complete system that has slowly been developed by gradual and persisting changes, that is, all present forms of plants and animals have arisen from pre-existing forms by a gradual process of change. 11. Fossil evidence and evidence from structural relationships, physiological relationships and the distribution of plants and animals. 12. Vol. II, Chap. 21. Evidence that Evolution has Occurred. 13. A character impressed on an individual by environmental influences and passed to succeeding generations. 14, 15. Vol. II, Chap. 21. Lamarckism. 16–19. Vol. II,

Chap. 21. Darwinism. 20, 21. Vol. II, Chap. 21. Weismannism. 22–25. Vol. II, Chap. 21. Neo-Darwinism.

Man (p. 87)

1. Genus and species, *Homo sapiens*; family, Hominidae; sub-order, Anthropoidea; order, Primates; class, Mammalia; sub-phylum, Craniata; phylum, Chordata. 2. *Australopithecus*; *Pithecanthropus*; *Sinanthropus* and possibly *Zinjanthropus*. 3. Man is fully bipedal; has characteristic dentition; a user of tools; has the power of speech; the big toe is non-opposable; possesses enlarged cerebral hemispheres with perfection of the neopallium. 4. By his ability to at least maintain his numbers; his status as the dominant living system on this planet; by his adaptability. 5, 6. Vol. II, Chap. 23. Man as a Successful Animal. 7–12. Vol. II, Chap. 23. Effects of Man on his Environment. 13. Dodo, moa and solitaire (flightless birds); Steller's sea cow. 14. Vol. II, Chap. 23. Effects of Man on his Environment. 15. Control of rabbit populations; destruction of cactus by *Cactoblastis*; eradication of citrus scale insect by ladybirds. 16. (1972) about 3500×10^6, increasing at a rate of about 30×10^6 per year. 17, 18. Vol. II, Chap. 23. Some Problems Affecting Man's Future Success. 19. Soil erosion; deforestation; depradation of fisheries; ineffective cultivation schemes; irreversible interference with natural populations; effects of pollutants. 20. Vol. II, Chap. 23. Some Problems Affecting Man's Future Success. 21. Venereal diseases and food poisoning increasing; smallpox and poliomyelitis decreasing. 22. The first two tend to spread more rapidly as human behaviour becomes less disciplined with respect to simple rules of personal and public health; the other diseases seem to be controlled by the conferment of immunity against them by the use of vaccines. 23. Genetical damage. The remaining questions, 24–28, should be answered from a general knowledge of human affairs.